NASA AERONAUTICS BOOK SERIES

Breaking the Mishap Chain

Human Factors Lessons Learned from Aerospace Accidents and Incidents in Research, Flight Test, and Development

Peter W. Merlin, Gregg A. Bendrick, and Dwight A. Holland

Published by Books Express Publishing
ISBN 978-1-78266-246-4

Books Express publications are available from all good retail and online booksellers. For
publishing proposals and direct ordering please contact us at: info@books-express.com

About the Authors

Peter W. Merlin is an aerospace historian with extensive knowledge of the various factors involved in aerospace mishaps. Under contract to the National Aeronautics and Space Administration (NASA) at Dryden Flight Research Center, Edwards, CA, since 1997, he has authored a variety of books, including several NASA Special Publications on aeronautical research projects. He served as coauthor of research pilot Donald Mallick's autobiography, *The Smell of Kerosene: A Test Pilot's Odyssey*, and *X-Plane Crashes: Exploring*  *Experimental, Rocket Plane, and Spycraft Incidents, Accidents and Crash Sites*, with Tony Moore. He has also authored several technical papers for the American Institute of Aeronautics and Astronautics, as well as numerous journal articles on aerospace history and technology. In addition, he serves as contributing editor for historical publications at Dryden and has appeared in more than a dozen documentary television programs for the Discovery Channel, the History Channel, National Geographic Channel, and others. He holds a bachelor of science degree in aviation management from Embry-Riddle Aeronautical University.

Gregg A. Bendrick, M.S., M.D., M.P.H., is the chief medical officer at NASA Dryden Flight Research Center. He previously served 9 years in the U.S. Air Force as a flight surgeon before joining the Ochsner Medical Center in New Orleans, LA, where he practiced occupational medicine for nearly 3 years. At Dryden he oversees all aspects of aerospace medicine, occupational medicine, and fitness center operations. Additionally, he is the medical review officer for workplace drug testing and served as onsite medical coordinator for  Space Shuttle landings at Edwards Air Force Base (AFB). He is board-certified in aerospace medicine and designated by the Federal Aviation Administration (FAA) as a senior aviation medical examiner. He is a fellow of the Aerospace

Medical Association, an instructor in the University of Southern California's Aviation Safety and Security Program, and an affiliate faculty member with Embry-Riddle Aeronautical University. He has authored several technical papers on various aspects of aerospace medicine as well as a novel.

Dwight A. Holland, M.S., M.S.E., Ph.D., M.D., is a principal with Human Factors Associates, a fellow in the Aerospace Medical Association, and an academician in the International Academy of Aviation and Space Medicine. He has served as the president of the International Association of Military Flight Surgeon Pilots and of the Space Medicine Association. In 2003, he served as the technical cochair of the largest International Systems Engineering Conference held to date. He has over 100 academic presentations, abstracts, book chapters, journal special editions, and papers to his credit. As a U.S. Air Force Reserve officer, he has taught and codeveloped the Human Factors in Flight Test courses at the Air Force and Navy test pilot schools. He holds commercial and jet ratings with over 2,000 hours' flying time in more than 40 civil and military aircraft. He also participated as a geophysicist on a remote field expedition to the Antarctic, later writing a study comparing the human factors issues of polar exploration to those of long-duration space flight operations.

Table of Contents

Acknowledgments   **vii**

Introduction   **ix**

Part 1: Design Factors

Chapter 1: "It May Not Be Hooked Up": Automation Bias, Poor Communication, and Crew Resource Management Factors in the X-31 Mishap **3**

Chapter 2: Habit Pattern Transfer During the First Flight of the M2-F2 Lifting Body **23**

Chapter 3: Pilot-Induced Oscillation During Space Shuttle Approach and Landing Tests **33**

Part 2: Physiological Factors

Chapter 4: Screening Versus Design: The X-15 Reentry Mishap **55**

Chapter 5: Six Million Dollar Man: The M2-F2 Task Saturation Mishap **81**

Chapter 6: Almost-Loss of Consciousness in the F-22A Raptor **101**

Part 3: Organizational Factors

Chapter 7: Decision Chain Leading to the XB-70/F-104 Midair Collision **127**

Chapter 8: Mission Management and Cockpit Resource Management in the B-1A Mishap **143**

Chapter 9: Collision in Space: Human Factors in the Mir-Progress Mishap **159**

Conclusions   **189**

Bibliography   **191**

Index   **207**

Acknowledgments

The authors would like to thank the many people who helped make this book possible. First of all, thanks to Tony Springer, NASA Aeronautics Research Mission Directorate, for sponsoring this project. We are grateful for the efforts of many people at NASA Dryden Flight Research Center including, but not limited to, Dr. Christian Gelzer, Tom Tschida, Karl Bender, and especially to Sarah Merlin for copyediting the final manuscript. Special thanks to Dr. Thomas Hoffmann and Dr. Philip J. Scarpa at the John F. Kennedy Space Center, Dr. William J. Tarver at the Lyndon B. Johnson Space Center, and Dr. James W. Butler, who reviewed our material for technical accuracy. Thanks to the staff of the Air Force Flight Test Center, particularly Dr. Craig Luther, for providing valuable source material and images. Apologies to anyone we missed. Any factual errors are the authors' responsibility. We made an attempt in good faith to get the facts straight by using the best available source material.

The popular image of test pilots as cowboys is misleading. Actual research pilots and test pilots like Joe Walker, seen here exiting the cockpit of the X-1A, were and are highly trained, educated professionals. (NASA)

Introduction

When a team of military and civilian researchers conquered the sound barrier with the rocket-powered Bell X-1 piloted by Capt. Charles E. "Chuck" Yeager in October 1947, flight performance was considered to be primarily a function of the airframe/powerplant configuration, flight controls, and pilot skill. Design focus in the 1940s and 1950s, sometimes called the golden age of flight test, concentrated on vehicle configuration; aerodynamic control problems; and performance during low-speed, transonic, and supersonic flight. Human-machine interaction was not considered a key issue, except for the narrow dimension of flying handling qualities.[1]

Within the flight-test community, a good pilot needed not only flying skills, but also engineering knowledge. An optimal pilot embodied a combination of the requisite skills, training, courage, and experience (i.e., the "right stuff," as described by author Tom Wolfe in his brilliant book of the same name).[2] Such pilots flew dozens of experimental craft, known collectively as X-planes, to evaluate a wide variety of cutting-edge configurations and capabilities. They flew in regimes of flight where new rules had to be written for aerodynamics, propulsion, navigation, and thermal effects.[3] Aviators were frequently exposed to extreme and unpredictable flight conditions that provided both physiological and cognitive challenges. In the early years of aviation, mishaps were often unfairly attributed to "pilot error," a catchall term too often used to describe a variety of issues related to human factors.

As early as the 1920s, however, a trio of Army Air Corps officers began to think about the effects of the human neurovestibular system on spatial orientation in flight. Bill Ocker, David Myers, and Carl Crane often went against the prevailing wisdom of the era, which was that instruments were not essential to flying safely in what are now termed "instrument flying conditions" (when visual references are obscured by darkness or weather). When Ocker and Crane

---

1. Richard P. Hallion, *Test Pilots: The Frontiersmen of Flight* (Garden City, NY: Doubleday, 1981); Peter W. Merlin and Tony Moore, *X-Plane Crashes—Exploring Secret, Experimental, and Rocket Plane Crash Sites* (North Branch, MN: Specialty Press, 2008).

2. Tom Wolfe, *The Right Stuff* (New York: Farrar, Straus and Giroux, 1979).

3. Dennis R. Jenkins, *X-15: Extending the Frontiers of Flight* (Washington, DC: NASA SP-2007-562, 2007).

wrote the very first book on instrument flight in 1932, *Blind Flight Guidance*, it became the basis for most aviation instrument-training programs of the day. The book was quickly adopted by the civilian aviation community and was used throughout the world but was only reluctantly accepted by the U.S. military.[4]

When James H. "Jimmy" Doolittle made the first successful airplane flight without the use of outside visual references—using only flight instruments—in 1929, those involved in aviation began to recognize the significance of the interaction between a pilot and an aircraft's controls and display systems. In the 1930s, as Sir Frederick Bartlett of Cambridge University began to research

pilot error in a simulator known as the Cambridge Cockpit, there was an increasing, though limited, realization that the design of the hardware interface could affect human performance.

During World War II, technological advances resulted in the production of faster, higher-flying, and more efficient aircraft. At the same time, researchers began to notice an increasing number of errors in human performance in the cockpit. As a scientific approach to flight operations replaced earlier, more intuitive methods, psychologists such as Paul Fitts began to write about the birth of modern human factors engineering.

By the 1950s and 1960s, it appeared as though aircraft designs were reaching the limit of human performance in terms of pilot workload and task saturation. Such aircraft as the F-4 Phantom and B-52 Stratofortress featured classic examples of steam gauge–type cockpits, with their confusing array of dials and switches. Pilots often found instrument faces too small to see clearly (especially during violent maneuvers), or

Jimmy Doolittle piloted the first airplane flight made without outside visual references—using only flight instruments—in 1929, demonstrating the significance of the interaction between a pilot and an aircraft's controls and display systems. (U.S. Air Force via Hill Air Force Base Museum)

---

4. Fred H. Previc and William R. Ercoline, eds., "Spatial Disorientation in Aviation," in *Progress in Astronautics and Aeronautics*, vol. 203 (Reston, VA: American Institute of Astronautics and Aeronautics, 2004).

Aircraft such as the B-52 Stratofortress featured cockpits with a confusing array of dials and switches. Pilots often found instrument faces too small to see clearly, and some instruments and switches were inconveniently located and difficult to reach. (NASA)

instruments and switches were inconveniently placed, sometimes hidden behind other equipment, and difficult to reach.

Modern "glass cockpits" feature electronic multifunction displays that a pilot can adjust and optimize according to need. Virtual instruments on these displays can be enlarged, reduced, or repositioned as necessary. Other displays can be called up as needed through multifunction keys or even voice commands.[5]

Although the introduction of computer technology into flight operations has reduced many problems and often increased pilot performance in the

---

5. Col. Art Tomassetti (United States Marine Corps [USMC]), "Flight Test 2040," presented at the 54th Annual Society of Experimental Test Pilots Symposium (Anaheim, CA, September 25, 2010).

Modern "glass cockpits," such as in the F-18 Hornet, feature electronic multifunction displays, which a pilot can adjust and optimize according to need. (NASA)

cockpit, human-computer interaction has also introduced a new source of error in complex systems. Lessons learned through close calls and mishaps have emphasized the need for a human factors design approach.[6]

Since its beginnings in the 1940s, the study of human factors (also known as ergonomics, engineering psychology, or human factors engineering) has gradually been broadened to include a variety of issues that affect human performance, as well as efficiency and safety.[7] Since the time of Yeager's first supersonic flight, repeated studies in a variety of situations have consistently demonstrated that despite state-of-the-art hardware and software, the interaction between human and machine remains an important variable. Accidents involving people, aircraft, and spacecraft, though once attributed simply to human error, have been shown to involve not only a wide range of human factors engineering issues, but also layers of operational, situational, and organizational factors.[8]

Thinking along these lines, human factors may be conceptualized as follows:

> The scientific discipline concerned with the understanding of interactions among humans and other elements of a system, and the profession that applies theory, principles, data, and other methods to design in order to optimize human well-being and overall system performance.[9]

---

6. John A. Wise, V. David Hopkin, and Daniel J. Garland, *Handbook of Aviation Human Factors* (Boca Raton, FL: CRC Press, Taylor and Francis Group, 2010).

7. Rashid L. Bashshur and Corinna E. Lathan, "Human Factors in Telemedicine," *Telemedicine Journal* 5, no. 2 (July 1999): 127–128.

8. Douglas A. Wiegmann and Scott A. Shappell, *A Human Error Approach to Aviation Accident Analysis: The Human Factors Analysis and Classification System* (Aldershot, U.K.: Ashgate Publishing, 2003).

9. Human Factors and Ergonomics Society, *http://www.hfes.org/web/AboutHFES/about.html*, accessed April 6, 2010. This definition was adopted by the International Ergonomics Association in August 2000.

Physical elements include such factors as cockpit design, human-machine interface, environmental constraints, and human physiological limitations. Psychological factors include but are not limited to mission planning, crew resource management (CRM), crew defensive psychological factors in place, individual/crew adaptability and flexibility, crew-controller interactions, and communications.

James Reason, professor emeritus at the University of Manchester, England, summarized the modern view of how human factors fits into a systems context with his "Swiss cheese" model of safety vulnerabilities in highly technical and complex organizations. His model incorporates not only the active errors that lead to a mishap, but also the latent errors (i.e., those conditions that are in place for some period of time before the active error occurs). These may include the immediate physical or environmental conditions surrounding the active error, as well as those practices, processes, and procedures that are further upstream in the organizational environment, such as supervisory standards, management directives, and corporate leadership. Each can set the conditions of vulnerability that, in one way or another, facilitate the propagation of an adverse event resulting from an active error. Reason described these conditions, or areas of vulnerability, as holes in the layers of defense established to guard against such error. These defenses include standard operating procedures, supervisory practices, managerial decisions, and corporate leadership.[10] The specific aviation mishap analysis taxonomy called the Human Factors Analysis and Classification System mentioned earlier has been developed based on Reason's Model and is currently in use within many aviation organizations.[11]

These holes of vulnerability and risk are not static. Over time, they do not remain the same size, nor do they remain in the same location within a particular layer of defense. In fact, they are constantly changing size, shape, and location as an organization, program, or project evolves over time, in ways related to the changing circumstances. For instance, upper-level management may change or technology may advance, introducing new variables. Though the changing size, shape, and position of the holes in the defensive layer are at times difficult to ascertain or predict, a total disregard for this process virtually guarantees a human factors–related failure, with an entire spectrum of potential consequences.

---

10. James Reason, *Human Error* (New York: Cambridge University Press, 1990).
11. Scott A. Shappell and Douglas A. Wiegmann, *The Human Factors Analysis and Classification System (HFACS)*, Report Number DOT/FAA/AM-00/7 (Washington, DC: FAA, Office of Aviation Medicine, 2000).

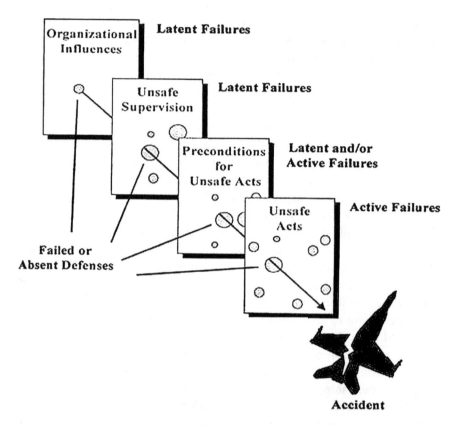

James Reason summarized the modern view of how human factors fit into a systems context with his "Swiss cheese" model of safety vulnerabilities in highly technical and complex organizations. Areas of vulnerability are "holes" in the layers of defense guarding against error. Accidents occur when holes align. (Author's collection)

This volume contains a collection of case studies of mishaps involving experimental aircraft, aerospace vehicles, and spacecraft in which human factors played a significant role. In all cases the engineers involved, the leaders and managers, and the operators (i.e., pilots and astronauts) were supremely qualified and by all accounts superior performers. Such accidents and incidents rarely resulted from a single cause but were the outcome of a chain of events in which altering at least one element might have prevented disaster. As such, this work is most certainly not an anthology of blame. It is offered as a learning tool so that future organizations, programs, and projects may not be destined to repeat the mistakes of the past. These lessons were learned at high material and personal costs and should not be lost to the pages of history.

Additionally, the book has been written in such a way as to be useful to a wide audience. Each case study includes a detailed analysis of aeromedical and organizational factors for the benefit of students, teachers, and others with an academic interest in human factors issues in the aerospace environment. For context, each story includes historical background that may be of more interest to general readers. The authors elected to include extensive biographical material on pilots and astronauts in order to highlight the fact that even the most qualified individuals can become links in the mishap chain.

Peter W. Merlin
Gregg A. Bendrick
Dwight A. Holland

Part 1:
Design Factors

The first X-31, seen here over Edwards Air Force Base, was built for the Enhanced Fighter Maneuverability demonstration program. A joint U.S.-German effort, it was the first international X-plane project. (NASA)

CHAPTER 1

"It May Not Be Hooked Up"

Automation Bias, Poor Communication, and Crew Resource Management Factors in the X-31 Mishap

In January 1995, a multimillion-dollar international experimental aircraft program was jeopardized by an avoidable mishap. The fortunately nonfatal accident happened as a result of several converging conditions, including several human factors. First, safety analysts relied too much on automated systems for alerting the pilot to airspeed data errors. Second, test team members failed to communicate the potential consequences of a configuration change that, under certain conditions, would result in such errors. Finally, inadequate CRM prevented important information from reaching the pilot in time, and a lack of situational awareness by the pilot added to the causal chain leading up to the mishap.

Enhanced Fighter Maneuverability

In air-to-air combat, survivability is largely dependent on maneuverability. As aerial combat evolved from the earliest experiences in World War I through the Vietnam conflict of the 1960s and 1970s, aircraft designers increasingly attempted to build more maneuverable aircraft.

In order to take advantage of improved aeronautical technologies, U.S. and allied military agencies conducted studies in the 1980s that led to a program called Enhanced Fighter Maneuverability (EFM). This resulted in the development of a highly maneuverable experimental aircraft called the X-31 with which to demonstrate agility and flight at unusually high angles of attack (AOA) (that is, the angle of an airplane's fuselage and wings relative to its flightpath). Through the use of thrust vectoring (directing engine exhaust flow) and various control surfaces, the X-31 could be safely flown at higher AOA than could conventional aircraft. Three thrust-vectoring paddles attached to the X-31's exhaust nozzle improved control, directing exhaust flow to provide control in both pitch and yaw. Additionally, two movable forward-mounted canards (small winglike structures) and two fixed aft strakes supplied additional control in tight maneuvering situations.

The primary goal of the X-31 research program was to provide aircraft designers with a better understanding of aerodynamics, the effectiveness of

Vectored thrust allowed the X-31 to be flown at extremely high angles of attack. An international team of pilots conducted military utility evaluations, pitting the X-31 against various fighter aircraft in simulated aerial combat. (NASA)

flight controls and thrust vectoring, and airflow phenomena at high AOA for use in the development of future high-performance aircraft. In mock combat demonstrations, the use of integrated flight controls, thrust vectoring, and other techniques that allowed maneuvering beyond normal flight envelope parameters provided the X-31 pilot with a tactical advantage. Additional program goals included the development of extremely short takeoff and landing capabilities as well as of semi-tailless configurations and advanced flight control systems (FCS).[1]

The EFM program constituted the first international effort to build an X-plane. The aircraft was designed and funded jointly by the U.S. Defense Advanced Research Projects Agency (DARPA) and the German Ministry of Defense. The design and construction of components and systems were shared between Rockwell International in the United States and Messerschmitt-Bölkow-Blohm (MBB, later Deutsche Aerospace) in the Federal Republic of Germany. Other U.S. participants included the Air Force, the Navy, and the National Aeronautics and Space Administration (NASA). The final assembly of two airframes took place at Rockwell's North American Aircraft Division plant in Palmdale, CA. The flight-test program included an international team of pilots from Rockwell, MBB, and the United States and German armed forces.

The X-31 was designed for subsonic flight only. Designers reduced overall program costs by using as many off-the-shelf components as possible. Airframe weight was reduced through the extensive use of graphite/epoxy thermoplastics in the construction of the aerodynamic surfaces and forward fuselage.

Assembly of the first X-31 A (Bu. No. 164584) was completed at Rockwell's Palmdale facility in February 1990, and Rockwell chief test pilot Norman K. "Ken" Dyson made the first flight on October 11, 1990. The first flight with the second aircraft was made January 19, 1991, with Deutsche Aerospace chief test pilot Dietrich Seeck at the controls.[2]

Following initial trials that included 108 sorties from Palmdale, flight operations moved to the NASA Dryden Flight Research Center at Edwards AFB. There, an international team of pilots and engineers expanded the aircraft's flight envelope and conducted military utility evaluations, pitting the X-31 against various fighter aircraft to evaluate its maneuverability in simulated aerial combat.

On September 18, 1992, the X-31 test team achieved a milestone with multiple demonstrations of controlled flight at 70 degrees AOA. A successful execution of a minimum-radius, 180-degree turn using a post-stall maneuver

---

1. Jay Miller, *The X-Planes: X-1 to X-45* (Hinckley, England: Midland Publishing, 2001).

2. Ibid.

in April 1993 proved the X-31 capable of exceeding the aerodynamic limits of any conventional aircraft.[3]

International Test Organization

The X-31 International Test Organization (ITO) was managed by DARPA and composed of representatives of both the United States and German Governments and private industry. A team at Dryden served as the responsible test organization for flight operations, aircraft maintenance, and research engineering.

The initial pilot cadre for the X-31 included Dyson and Fred Knox of Rockwell International, Dietrich Seeck, and German Ministry of Defence test pilot Karl-Heinz Lang. Eventually, 14 U.S. and German pilots would fly the X-31 a total of 580 times during the EFM program.[4]

Lang, who flew 116 flights between March 1991 and January 1995, was an early standout for a variety of reasons, not least of which was his habit of singing while conducting test maneuvers. But he was in all ways a professional in the cockpit.

German test pilot Karl-Heinz Lang flew 116 flights in the X-31. He was known for his habit of singing while conducting test maneuvers, but he was a consummate professional in the cockpit. (NASA)

Lang completed X-31 ground school training on March 8, 1991, and flew his first familiarization flight one week later. In July 1991, shortly after accomplishing his proficiency check flight, Lang also qualified as an X-31 instructor and flight examiner. He was also qualified as a T-38 pilot and held a German military pilot's license with instrument flight rules (IFR) rating and a Class 1 experimental flight rating, as well as a Federal Aviation Administration (FAA) commercial pilot's certificate (IFR/multiengine). By January 1995, Lang had accrued 4,775.9 flight hours in more than a dozen types of aircraft including the PA-200 Tornado, F-104G, F-8, F-16, F-18, Alpha Jet, T-38, X-31, and others.[5]

In a September 1994 memo to X-31 deputy program manager Helmut Richter, X-31 ITO director and NASA project manager Gary Trippensee wrote: "Lang's positive

---

3. Ibid.

4. Richard P. Hallion and Michael H. Gorn, *On the Frontier: Experimental Research at NASA Dryden* (Washington, DC: Smithsonian Books, 2003).

5. Dryden Flight Research Center (DFRC), NASA 584 X-31 Mishap Investigation Report, August 18, 1995, DFRC Historical Reference Collection, Edwards, CA.

attitude and professional approach to his X-31 test flying have been an instru-
mental ingredient in the success of the program to date."[6]

For his contributions to the program, Lang was awarded the NASA Public
Service Medal in 1994. An accompanying citation noted that he had provided
"consistent, strong, professional leadership" in efforts to achieve program goals.[7]

No Pitot Heat

On January 18, 1995, Lang flew
the first X-31 on its 289th sortie.
Instrumentation engineers had
installed a flutter-test mechanism
designed to allow accurate estimation of aerodynamic parameters. A Kiel probe was mounted
on the airplane's nose to collect
airspeed data. Unlike a conventional pitot tube, the Kiel probe
had a shrouded tip for more accurate measurements at high AOA.
Although the geometry of the Kiel
probe made it more susceptible to
icing, it had been used on the previous 150 flights without incident.

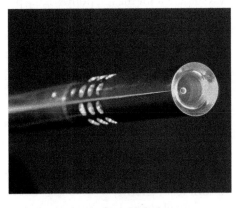

A Kiel probe was mounted on the X-31's nose to
collect airspeed data. The shrouded tip permitted
more accurate measurements at high angles of
attack but was highly susceptible to icing. (NASA)

The aircraft's final three flights were scheduled for the following day. It was then to
be subject to maintenance and possible mothballing, as program funds were low.
Meanwhile the second airframe had already been placed in storage at Dryden.[8]

On the morning of January 19, the test team held a preflight briefing to
prepare for two sorties to demonstrate a quasi-tailless configuration (piloted
by Gus Loria and Quirin Kim) and one final parameter-identification flight
by Lang. Since the maneuvers were essentially identical to those flown the
previous day, the test conductor presented an abbreviated control-room procedures briefing. Loria briefed procedures for the first two sorties, and Lang
then briefed his mission.

---

6. X-31 ITO memo dated September 6, 1994, X-31 correspondence files—personnel, DFRC
Historical Reference Collection, NASA Dryden Flight Research Center, Edwards, CA.

7. Nomination and citation documents, NASA Public Service Medal for Karl-Heinz Lang, X-31 correspondence files—personnel, DFRC Historical Reference Collection.

8. Michael A. Dornheim, "X-31 Board Cites Safety Analyses, but Not All Agree," *Aviation Week &
Space Technology* (December 4, 1995): 81–86.

The first two flights proceeded as planned. In the early afternoon, Lang and the test conductor discussed procedures for the final flight and worked with engineers to revise the test altitude based on forecasted cloud ceilings. Although the original plan called for taking data points at an altitude of 28,000 feet above sea level, weather observations indicated that the sky would be overcast between 23,000 and 25,000 feet. Having agreed that any altitude above 20,000 feet was sufficient for the test points, Lang proceeded to the airplane while control-room personnel received a briefing on the revised plan.[9]

Following takeoff, Lang climbed to altitude accompanied by Dana Purifoy in a NASA F-18 chase plane. All went according to plan, and Lang performed his test points at altitudes between 20,000 and 24,000 feet. He noted that there was a cirrus cloud deck at 23,000 feet and that he had no clear view of the horizon. Throughout much of the mission, the wingtip vortices of the X-31 generated visible condensation vapor. Recognizing that the abundance of water vapor in the air could lead to icing in the airplane's pitot tube (nose-mounted air data probe), Lang informed the test conductor that he had turned on the pitot heating switch and asked to be reminded to turn it off later.

Acknowledging Lang's transmission, the test conductor heard an engineer in the control room comment that pitot heating had not been hooked up for the X-31's Kiel probe. Following the test point then in progress, he asked for clarification. The engineer repeated that "Kiel probes don't have pitot heat," but this exchange was unheard by the pilot, who was setting up the next test point and whose radio, in any case, was tuned to a different frequency.[10]

As he prepared for his final test, Lang noted that his airspeed was off nominal, indicating 207 knots at 20 degrees AOA (which was impossible, given the power settings and AOA). He suspected pitot icing but assumed the heating system was functional. After completing the test point, Lang and the test conductor began going through the prelanding checklist. When the test conductor reminded the pilot to turn off his pitot heat switch, Lang said: "I think I'll leave it on for a moment."[11]

The test conductor, recalling the engineer's earlier comment, informed Lang: "Yeah, we think it may not be hooked up." Sarcasm fairly dripped from Lang's voice as he responded: "It may not be hooked up. That's good. I like that." Four seconds later he heard a warning tone in the cockpit and exclaimed: "Oh, God."[12]

---

9. DFRC, Mishap Investigation Report, NASA 584 X-31.

10. Ibid.

11. Ibid.

12. Ibid.

The X-31 bucked wildly in a rapidly increasing series of pitch oscillations until the nose was about 20 degrees past vertical, followed by a sharp roll. Recognizing that there was no hope of recovery, Lang ejected. He parachuted to safety but sustained serious back injuries. The X-31 crashed in a sparsely populated area less than 2 miles west of the community of North Edwards, CA, and just north of a highway. Fortunately, there was no damage to private property.[13]

Heating Elements and Human Elements

An accident investigation quickly focused on the pitot-static system that provided air data to instruments in the cockpit, the aircraft's flight control computers (FCC), and the mission control monitors at Dryden. The original aircraft configuration had included a Rosemount probe equipped with a heating element to prevent icing. It had been replaced later in the program, however, with an unheated Kiel probe that provided more accurate measurements at high AOA. The investigation revealed that partial pitot icing in the Kiel probe resulted in an indicated airspeed significantly lower than the actual airspeed and that the fly-by-wire system responded with excessive control gains that caused the aircraft to become unstable.[14]

Further investigation revealed that a number of human factors were causal or contributory to the accident. The primary factor in this regard was a lack of appreciation among test personnel for the potential consequences of losing accurate pitot-static pressure data. As a result, test planners failed to rigorously follow standard operating procedures to reduce this risk factor. Among the early consequences of the neglected procedures was a system-safety analysis that incorrectly assumed that faulty air data from the pitot-static system would be annunciated by the FCS, alerting the pilot to the problem. Additionally, project personnel demonstrated a general lack of awareness of the Kiel probe configuration with regard to the need for, or lack of, a heating element. Mission planners never intended for the airplane to be flown in conditions conducive to icing and so never saw a significant need for anti-icing systems. Finally, ineffective communication among mission control personnel prevented critical information regarding the lack of pitot heating from reaching the test conductor and pilot before it was too late. Had such information been available before the X-31 departed controlled flight, the pilot could have saved the airplane by pushing a single button.[15]

---

13. Ibid.
14. Ibid.
15. Ibid.

The flight control system panel is located in the center of the X-31 cockpit, below the head-up display. The FCS operated in several preprogrammed modes, depending on flight-profile parameters. These included a basic mode, three reversionary modes, and a spin-recovery mode. (NASA)

Automation Bias

The X-31 was a fly-by-wire airplane that was, in essence, flown by the computer based on input from the pilot. Specifically, the FCS consisted of 4 Honeywell flight-control computers and 10 actuators associated with the various aerodynamic and thrust-vectoring control effectors. Critical inputs to the FCS included, among other things, air data from the pitot-static system through two air data computers. In order to maximize maneuverability, the FCS adjusted the control gains—greater or lesser movement of the aircraft's aerodynamic control surfaces—based on aircraft speed and orientation.

The FCS operated in several preprogrammed modes, depending on flight profile parameters. These included a basic mode, three reversionary modes, and a spin-recovery mode. In the basic mode, FCS gains were scheduled according to Mach number, pressure altitude, and AOA. The three reversionary (R) modes were designed to handle specific types of failures. Mode R1 could be activated in response to Inertial Navigation Unit failure. Mode R2 was a response to flow-angle failures (that is, AOA and angle of sideslip), and mode R3 was to be used for air-data failures. The pilot could select each mode by pressing a button on the cockpit control console.[16]

FCS reliability was based on the principle of redundancy management. Air data computer (ADC) no. 1 provided both total pressure and static pressure to FCCs nos. 1 and 2, while ADC no. 2 fed to FCC no. 3. All three computers then independently performed control-law calculations, and each sent command signals to the various actuators. FCC no. 4 functioned as the tiebreaker among the three FCCs and the two ADCs.

If the FCS determined that a reversionary mode was needed, the post-stall thrust-vectoring feature was disabled, if active. A warning tone then sounded in the pilot's headset, and the appropriate reversionary mode switch would illuminate and flash. The pilot would manually activate the appropriate reversionary mode, as the system was not designed to automatically select one without pilot

---

16. Ibid.

input. In the case of mode R3, the FCSes selected a fixed set of parameters that provided stable operation (without available air data) over a wide operating range, including the landing phase. In fact, early in the program, ADC failures occurred on several occasions, but the pilot was able to land the aircraft uneventfully in mode R3.[17]

A system-safety analysis originally performed by the contractor in 1989 correctly identified a potentially hazardous condition in which an airspeed failure could result from a loss of pitot-static signal. The consequences were determined to be critical (one category below the worst, catastrophic). Moreover, the analysis identified a range of possible causes, including a plugged pitot tube. However, the consequences of this hazard were listed as "Local Effect—System switches to reversionary mode" and "End Effect—Decrease flight control performance." A later subsystem-safety analysis likewise concluded that mode R3 would mitigate the hazard, resulting only in the degradation of FCS performance. A subsequent 1992 NASA study surmised that airspeed failure would result in "degraded flight control performance" and therefore found it to be an "accepted risk" (one category below "critical"). In other words, the hazard controls in the various system-safety analyses cited the availability of the reversionary modes as having completely addressed the potential hazard. Although the critical importance of valid data from the ADCs was noted, the 1992 report asserted that the ADCs could reliably be expected to identify invalid air data. Unfortunately, this conclusion was simply incorrect.[18]

The FCS could identify an air-data failure only if the two ADCs disagreed. Although airspeed-indication errors could be detected through differences between data provided by the two ADCs, both were fed by a single pneumatic source. If flawed information from this common source led to gradual degradation of air-data quality rather than a precipitous disruption, it would not be detected because the FCS was simply not designed to do so. The X-31 mishap investigation revealed a latent misconception among many test team members and system-safety engineers that the FCS actually did have this capacity. Therefore, no operating or emergency procedures were written specifically for responding to a gradual disruption of the single air-data source. According to the Accident Investigation Board:

> The fallacy throughout these documents was the misconception that the FCS was capable of detecting the full range of pitot-static failures. As discussed in the FCS and pitot-static sections above,

---

17. Ibid.

18. Ibid.

the FCS was solely capable of detecting errors arising because of disparities in the output of the ADCs and was incapable of determining the validity of the single-string pneumatic signal entering the ADCs.[19]

During a lengthy hardware-in-the-loop simulation effort conducted in preparation for the aircraft's first flight, failures in the total and static pressures were inputted into the system with the aircraft configured for a variety of flight conditions. In these instances, pitch departures occurred, resulting in the simulated loss of the aircraft. However, a "System Problem Report and Disposition" (a Rockwell problem-reporting and corrective-action system) was not initiated because engineers concluded that such a failure conceivably could occur only as the result of a bird strike or some other such precipitous loss of the pitot boom during either takeoff or landing. Engineers calculated that the aircraft would remain controllable long enough to allow the pilot to select mode R3. Unfortunately, this conclusion was not verified or coordinated with responsible Dryden flight operations personnel. In any case, icing of the pitot tube was not considered a potential causative factor for air-data loss because flight operations of the aircraft were restricted to visual meteorological conditions, which prohibited flight into icing conditions. Likewise, the aircraft's original configuration, with the Rosemount probe, included functional pitot heat. In short, the failure modes and effects identified during the simulation analysis were thought to be implausible.

More ominously, the simulation sessions clearly identified an earlier system-safety analysis that had contained significant errors, but that report had never been modified or revisited. The absence of documentation regarding the problem virtually guaranteed that future project participants would remain unaware of this identified hazard.[20]

As investigators noted:

> Flawed analysis of the design of the pitot-static system obscured a latent severe hazard to safe operation of the X-31 aircraft. This in turn fostered a situation in which all of the pilots, engineers, management, and government oversight associated with the program placed undue confidence in the ability of the redundancy management system to detect flight critical faults.[21]

---

19. Ibid., section 7.4.1, pp. 7–11.
20. Ibid., section 7.4.1, pp. 7–13.
21. Ibid., section 7.4.4, pp. 7–16.

Human factors engineers refer to the tendency to rely on computerized hardware and software to the detriment of accurate analysis and decision making as "automation bias." The term is normally applied to a single individual's focused attention on an instrument display or a programmed action to the exclusion of other information that would reveal the automated action to be incorrect. Yet it seems in the case of the multiple system-safety analyses for the X-31 program that automation bias affected all engineers and managers involved. Recognition of this fact, along with some justified cynicism and pointed questioning of the system-safety analyses during numerous reviews—particularly in light of the simulation studies—might have led to a more accurate systems-safety analysis for this program.[22]

Lack of Configuration Awareness

As previously noted, the original X-31 system-safety analysis did not fully consider pitot icing as a hazard because the aircraft would be operated only in visual meteorological conditions. Likewise, the aircraft was originally equipped with a probe configuration that included a pitot heating system. However, with the introduction of the Kiel probe, two factors were introduced that raised the risk of failure and consequent loss of valid air data. The first of these was the probe's susceptibility to icing under conditions that would not cause icing in a standard pitot system, such as the original Rosemount probe. The second was the absence of pitot heating in the Kiel configuration.

The Kiel probe had a curved tip at the end to facilitate airflow at high AOA and a Venturi geometry that made it more susceptible to icing. As air entered the probe, pressure decreased such that with the right combination of temperature and humidity, ice would form under conditions where icing was not otherwise a concern for flight. Although obvious in retrospect, these design aspects were not fully appreciated prior to the mishap. Investigators noted:

> There appears to have been no *a priori* knowledge of the Kiel probe's susceptibility to icing.... While the icing susceptibility can be deduced by inspection, given knowledge of the mishap, it is apparent that no one associated either with the probe or the ITO [International Test Organization] imagined that this mishap might be an outcome of its use.[23]

---

22. Christopher D. Wickens, John D. Lee, Yili Liu, and Sallie E. Gordon Becker, *An Introduction to Human Factors Engineering*, 2nd ed. (Upper Saddle River, NJ: Pearson Prentice Hall, 2004).

23. DFRC, NASA 584 X-31 Mishap Investigation Report, section 7.4.3, pp. 7–15.

Moreover, X-31 flight operations in icing conditions were prohibited. It should be noted, however, that the prohibited conditions were those in which ice would typically develop on the wings of the aircraft, thereby disrupting aerodynamic flow and resulting in consequent loss of lift. Such conditions usually involve visible precipitation. Icing of the pitot tube itself is ordinarily not a concern since pitot heat is usually available. Conditions that would lead to icing of the Kiel probe were not common in the vicinity of Edwards AFB, located on the western edge of the Mojave Desert in Southern California.

The X-31 had been flown in this configuration about 150 times without incident. Shortly before the mishap, the pilot—recognizing the environmental conditions from previous flight experience—correctly identified pitot icing as the likely cause of the airspeed discrepancy he had observed. He activated the pitot heat switch in the cockpit, asking the NASA test conductor to remind him to turn it off later. He was clearly unaware that the configuration of this aircraft, with its use of the Kiel probe, had no pitot heating mechanism.

The configuration control process that should have prevented this problem had several deficiencies. The first, called the Temporary Operating Procedure (TOP) system, was a Rockwell process adopted by the ITO at the program's outset to convey changes in operating procedures. The original TOP for installation and operation of the Kiel probe was drafted by a flight-controls engineer and would have informed all project personnel that the Kiel probe was unheated. But this TOP was somehow lost during the publishing process. There was no formal closed-loop tracking procedure to ensure that such changes were incorporated prior to publication, nor was there any formal training required for changes that were adopted. As a result, this missing document (known as TOP no. 7) was never distributed. Though it was eventually found in the memory of the originator's computer, precisely how it was misplaced was never determined. Subsequent investigation revealed that at the time of the mishap, four of the five active X-31 test pilots believed that a pitot heating system was still operative after the Kiel probe had been installed.[24]

A second deficiency of the configuration control process involved the mini tech brief provided by project personnel following a configuration change or a change in operating procedure. This technical briefing was presented to Dryden management on April 6, 1993, during a preflight meeting in order to obtain approval to fly with the Kiel probe. However, there was no written reference in the notes from this meeting addressing the fact that the pitot heat had been disconnected or the potential risk this implied. The air-data engineer who presented the briefing later stated that the inoperative pitot heat system had been

---

24. DFRC, NASA 584 X-31 Mishap Investigation Report.

discussed by attendees. The conclusions of that discussion, however, were not documented. Due to the inherently dry environmental conditions prevalent at Edwards AFB, the likelihood of increased risk apparently was discounted.[25]

Third, a cockpit placard informing the pilot as to the status of the pitot heating mechanism (e.g., "pitot heat inoperative") was never installed in the aircraft. Project personnel missed several opportunities in the work-order process to address this need. The first was in the Configuration Change Request for the new Kiel probe, which made no mention of pitot heating. The author of a subsequent engineering change order simply assumed the probe was heated. The final work order, issued for the installation of the Kiel probe, correctly specified the circuit breaker for the pitot heat to be "collared off" to prevent it from being pushed. It did not specify, however, any requirement that the pitot heat switch in the cockpit be placarded as inoperative, as is standard practice in the aviation industry for cockpit switches that are not functional.[26]

Throughout the X-31 program, the airplane's pitot probe configuration underwent changes according to the need to collect many different datasets. Sometimes the aircraft was flown with the heated Rosemount probe and sometimes with the Kiel probe. Depending on the planned flight profile, the maintenance crew chief installed or removed the probe in accordance with a work order, sometimes switching the Rosemount with the Kiel between multiple flights during the course of a single day.[27]

During the X-31 mishap investigation, the crew chief stated that he told the pilot in the preflight briefing that the pitot heat was "Inop[erative]." Lang, however, denied receiving this information. In any case, there had been a change order put through to include "pitot heat?" as a pilot's preflight checklist item, but this change had not yet been implemented at the time of the mishap.[28]

One other deficiency in the configuration-change management process for the X-31, which was not noted in the accident report, was the lack of any sort of human factors engineering review in such configuration changes. While

---

25. Ibid.

26. FAA, Office of the Chief Scientific and Technical Advisor for Human Factors, *Guidelines for Human Factors Requirements Development*, AAR-100, ver. 1.0, February 6, 2003; FAA Advisory Circular AC No. 120-51E, "Crew Resource Management Training," January 22, 2004; D. Wagner, J.A. Birt, M. Snyder, and J.P. Duncanson, *Human Factors Design Guide for Acquisition of Commercial-Off-The-Shelf Subsystems, Non-Developmental Items, and Developmental Systems*, DOT/FAA/CT-96/1, National Technical Information Service, January 1996.

27. Robert Cummings, personal interview by Gregg Bendrick at NASA Dryden Flight Research Center, March 17, 2004.

28. Ibid.

experimental aircraft flown under the auspices of NASA are not required to undergo formal certification by the FAA—and hence are not necessarily required to follow FAA rules regarding configuration changes and aircraft modifications—it is still incumbent upon aircraft designers to ensure that fundamental principles of aircraft and cockpit design are followed. The field of human factors engineering places a great deal of emphasis on the design and location of various displays and controls within an aircraft, and specifically on the cockpit's instrument panel. Had there been even the most cursory human factors engineering review of the configuration change for the Kiel probe, it is extremely doubtful that such a change would have been allowed to proceed without a mandatory placement of a cockpit placard indicating that the pitot heating system was inoperative.

Crew Resource Management

Mission control-room communications and CRM supplied the final human factors contributing to the mishap. The pilot was familiar with the type of environmental conditions in which icing is prevalent. He recognized the possibility that such conditions were the likely cause of the instrument discrepancies and activated the toggle switch to engage a pitot heating system he believed to be functional. A coincidental airspeed decrease—which data later suggested was due to aircraft descent into a zone of higher static pressure—gave him a false assurance that the heat was working.

When Lang informed the NASA test conductor that he had turned the pitot heat on and asked to be reminded to turn it off later, the test conductor acknowledged that request without apparent knowledge that it was inoperative. He was alerted to the fact by the team's test engineer but, distracted with accomplishing a test point, did not immediately inform the pilot. Upon finally being informed that his attempts to address the icing problem had been in vain, Lang responded with obvious frustration. Moments later, the airplane departed controlled flight and the final opportunity to avert disaster had been lost. At any time prior to loss of control, he could have activated mode R3.

The Mishap Investigation Board did not fault the pilot for failing to realize that the pitot heat was not functioning because analysis indicated pilot actions were based on the fact that Lang lacked sufficient information about the inability to heat the Kiel probe.[29]

Basically, the shortcomings of configuration control processes led to a situation in which the pilot logically but erroneously thought he had functional

---

29. DFRC, NASA 584 X-31 Mishap Investigation Report, section 7.5.4.2, pp. 7–21.

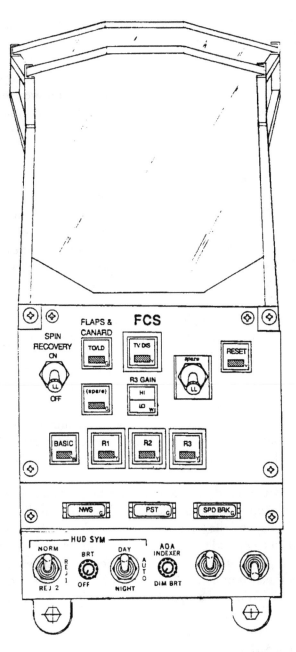

Flight control system gains were scheduled according to Mach number, pressure altitude, and angle of attack. Three buttons at the bottom of the FCS control panel activated reversionary modes designed to address specific types of failures. Mode R1 was used for inertial navigation unit failure, mode R2 was a response to flow-angle failures, and mode R3 was to be used for air-data failures. (NASA)

pitot heat. This assumption conceivably facilitated continuing the mission into less-than-ideal weather conditions.

The Mishap Investigation Board likewise concluded that the actions of mission control personnel were not causal to this mishap, taking into account their overall training, knowledge, and experience.[30] However, there are some aspects of communication, aviation psychology, and group dynamics that are applicable to this situation and might have been used to improve overall group awareness in the situation as it unfolded. For example:

> The control room was not prepared to monitor, recognize, and respond to the degradation of flight critical Pitot-static param-eters.... [T]he failure of the mishap pilot to report his first erro-neous airspeed indication and the failure of the control room to act and call for a pause in the mission to understand the mishap pilot's airspeed error calls, does not reflect the overall test team's philosophy as briefed to the Board after the mishap.[31]

The test team espoused a philosophy under which anyone who had input could speak freely at any time. However, it appears from the transcript of control-room communications that there was some hesitancy in this regard, perhaps due to excessive deference to authority and/or a perception that a junior team member's input would not be taken seriously by those more senior.[32]

In any event, the fundamental failure in this case was the fact that someone in the control room had a proper awareness of the situation, yet this critical piece of information was not effectively communicated to the individual, that is, the pilot, who was in a position to act on it. This was by definition a com-munication failure.

Fatigue and, to some degree, complacency likewise may have been factors in such a lack of optimal communication. This flight was the third made on that particular day and the seventh in 2 days. It also involved a relatively low workload when compared to the two previous missions. It was to be the last mission of the program for this particular aircraft, which was then going to be either mothballed or completely refitted for a new project. The mishap occurred on the return-to-base portion of the flight profile, with the mission

---

30. Ibid., section 7.5.6.2.

31. Ibid.

32. H. Clayton Foushee and Robert L. Helmreich, "Group Interaction and Flight Crew Performance," in *Human Factors in Aviation*, ed. Earl L. Weiner and David C. Nagel (San Diego, CA: Academic Press, 1988) pp. 189–227.

The X-31 crashed in sparsely populated desert terrain within a quarter mile of an occupied home and a major highway. (NASA)

nearly finished. Human performance in these situations is such that faulty situational awareness, ambiguous statements, and the lack of forceful reiteration in communications are not uncommon. Furthermore, the indicated values for airspeed, AOA, and engine power setting were totally inconsistent as the aircraft descended during the return to base. The fact that the pilot failed to notice this indicated his lack of global situational awareness since those airspeeds and power settings at those AOA were simply not possible. Ideally, the pilot should have questioned controllers about this, and control-room personnel also should have caught the discrepancy during the descent phase of the mission.

Lack of situational awareness and faulty communication have led to numerous aircraft accidents over the years, most notably the crash of two 747s on the island of Tenerife, Spain, in 1977.[33] Since that time, various governing

---

33. Steven Cushing, *Fatal Words: Communication Clashes and Aircraft Crashes* (Chicago, IL: University of Chicago Press, 1997).

bodies and research organizations associated with aviation, using the principles of social psychology and organizational behavior, have expended a great deal of effort exploring how individuals function within small groups. These efforts have led to the development of the principles of CRM, which have since become standard within the realm of aviation.[34] In its broadest sense, CRM is the use of all available resources, information, equipment, and personnel to achieve a safe and efficient flight. It focuses on how individuals develop the mental model of a situation and how they modify or reinforce these mental models by communicating with others, recognizing the potential for fatigue and other human performance issues to interfere with these processes. Since its introduction at a NASA-sponsored workshop in 1979, the CRM concept has evolved through several generations,[35] and while initial efforts have focused on operational aircrew and air traffic controllers,[36] these same principles have since been applied to other settings, such as aircraft maintenance, the nuclear power industry, and even healthcare.[37] Though the contexts and settings may change, the variable of human performance is the lone constant.

Conclusions

The X-31 mishap of January 19, 1995, like all accidents, had multiple contributors, many of which fall under the category of human factors. The first broad area in this regard was the error in thinking that the vehicle's FCS would identify and annunciate errors in airspeed data resulting from an ice-clogged pitot tube. In this sense there was an automation bias on the part of those who produced the system-safety analysis.

The second human factors error was in the failure to appreciate the role of pitot heat—and the Kiel probe's lack thereof—among all test team members. Of primary importance in this regard, from the perspective of human factors engineering, was the failure to placard the cockpit pitot heat switch as inoperative and to properly alert the pilot to the inconsistent airspeeds for AOA and power settings. In engineering, configuration control is critical. Changes to

---

34. FAA Advisory Circular AC No. 120-51E, "Crew Resource Management Training," January 22, 2004.

35. R.L. Helmreich, A. C. Merritt, and J. A. Wilhelm, "The Evolution of Crew Resource Management Training," *International Journal of Aviation Psychology* 9, no. 1 (1999): 19–32.

36. FAA Advisory Circular AC No. 120-72, "Maintenance Resource Management Training," September 28, 2000.

37. Robert L. Helmreich and Ashleigh C. Merritt, *Culture at Work in Aviation and Medicine: National, Organizational and Professional Influences* (Farnham, Surrey, U.K.: Ashgate Publishing Ltd., 2001).

even a small part of a complex system like that of a research airplane can have significant consequences. The details must be clearly communicated to affected personnel any time configuration changes are made.

The third failure involved lack of appropriate CRM among control-room personnel, such that the person with the critical piece of information regarding the status of the air-data collection system was unable to effectively communicate this information to the individual in a position to act upon it. The value of formal incorporation of human factors engineering review in the configuration-change management process and the value of formal, repeated training in the principles of CRM by all test team members, including control-room personnel, are among the human factors lessons to be learned from the crash of the X-31 enhanced fighter maneuverability demonstrator.

The last partial failure occurred when the pilot failed to recognize and, consequently, did not verbalize that something was wrong; the air-data computer output was giving the pilot information via his head-up display about airspeed, power settings, and AOA that simply was not possible. Flight-test profiles are often crammed with many subtle and not-so-subtle details that a pilot must be aware of and that contribute to high workload. In this case the pilot's awareness of the aircraft's overall energy state relative to expected outputs indicated on his flight parameter displays was poor. Additionally, the aircraft's design configuration was of little help in making this aspect of the mission more manageable.

The M2-F2 was the first heavyweight wingless lifting body research aircraft. The half-cone configuration was flown to demonstrate potential characteristics for a future space vehicle capable of entering Earth's atmosphere and landing on a conventional runway. (NASA)

CHAPTER 2

Habit Pattern Transfer During the First Flight of the M2-F2 Lifting Body

On July 12, 1966, a team of NASA and Air Force personnel conducted the first flight of a heavyweight experimental wingless lifting body vehicle at Edwards AFB. This was the culmination of an effort to develop a technique for landing a piloted spacecraft on a runway following atmospheric entry from orbit. The events of that flight also provided an example of human factors difficulties resulting from negative transfer of experience and training.

Flying Without Wings

Early piloted spacecraft designs featured ballistic entry into the atmosphere, similar to that of a missile warhead. This method resulted in high g-loads and intense heating due to atmospheric friction. Ground-test subjects in a centrifuge demonstrated that humans tolerated extended periods of high g-loads better when force was applied from front to back, the so-called "eyeballs-in" mode. Hence, Mercury (the first U.S. piloted space vehicle and the only one to use a strictly ballistic entry trajectory) was designed so a single crewmember could lie on his back facing away from the direction of flight. Final deceleration was accomplished through the use of a parachute, and the capsule landed in the ocean.[1]

During the Gemini and Apollo programs, a semiballistic entry provided a small amount of lift (aerodynamic force perpendicular to the flightpath) during entry. This allowed for increased cross range and more accurate guidance to the recovery point, as well as decreased g-loads and temperatures. Aft-facing, semireclined crew position and parachute recovery were still necessary, leading some designers to pursue a configuration known as a lifting body that could be flown much like a conventional airplane to a controlled landing on a runway.

---

1. Robert Hoey, "Testing Lifting Bodies at Edwards," in *Air Force/NASA Lifting Body Legacy History Project* (Lancaster, CA: PAT Projects, Inc.,1994), pp. 1–9.

The defining characteristic of such a vehicle was a fully lifting entry into Earth's atmosphere during which the pilot could adjust the flightpath continuously, changing both vertical motion and flight direction while decelerating from orbital velocity to a safe landing speed. The vehicle configuration allowed the pilot to be seated facing forward as in a conventional airplane cockpit. Cross range was maximized while g-loading and heating (despite longer entry times) were reduced. The most favored lifting entry configurations were delta-winged craft (such as the X-20 and Space Shuttle orbiter) and wingless lifting bodies.[2]

In developing the latter, National Advisory Committee for Aeronautics (NACA) engineers H. Julian Allen and Alfred Eggers of the Ames Aeronautical Laboratory, Moffett Field, CA, began studying blunt-body aerodynamics in 1950. A blunt cone shape produced a strong detached bow shock that provided thermal protection but had an extremely low (less than 1.0) lift-to-drag ratio (L/D) and experienced 8 g's reentry loads (greater than comfortably acceptable for human crew). Allen and Eggers joined with fellow engineers Clarence Syvertson, George Edwards, and George Kenyon to design a blunt half-cone configuration for a piloted orbital vehicle with a much more acceptable lift-to-drag ratio of around 1.5, resulting in a 1-g acceleration load and a cross range of more than 1,500 miles from the initial point of atmospheric entry.

By 1960, the Ames team had refined the configuration to a 13-degree half-cone with a blunted nose and a hypersonic L/D of 1.4. Inadequate subsonic stability necessitated tapering of the aft end and the addition of vertical stabilizer fins. The basic configuration, designated M2, served as the basis for a series of subscale radio-controlled models, a lightweight piloted glider, and a heavyweight rocket-powered vehicle.[3]

Robert Dale Reed, an engineer at NASA's Flight Research Center (now known as the Dryden Flight Research Center) at Edwards AFB, became interested in potential applications for lifting bodies. In 1962 he spoke with researchers at Ames and other NASA Centers who expressed skepticism that such a design could be built without some sort of deployable wings or pop-out engines to enable a safe landing. Experiments with scale M2 models, however, gave him confidence that a full-scale piloted craft could be flown safely, and he soon enlisted other engineers in his cause. They approached Flight Research Center Director Paul F. Bikle with a proposal for a full-scale

---

2. Ibid.

3. Richard P. Hallion and Michael H. Gorn, *On the Frontier: Experimental Flight at NASA Dryden* (Washington, DC: Smithsonian Books, 2003), pp. 143–171.

piloted glider version of the M2 that would be towed aloft behind an airplane. An endorsement from research pilot Milton O. Thompson sealed the deal.[4]

Engineering Research Pilot

Thompson had trained as a naval aviator near the end of World War II and served several years as a fighter pilot. While flying in a Naval Reserve squadron in Seattle, WA, he worked for Boeing as a structural-test engineer and flight-test engineer and earned a degree in engineering from the University of Washington. Thompson joined the NACA at Edwards AFB in 1956, at what was then known as the High-Speed Flight Station. After working as a research engineer for 2 years, he transferred to a position as research pilot in such aircraft as the F-51, T-33, F5D-1, F-100, F-101, F-102, F-104, F-105,

NASA research pilot Milton O. Thompson prepares to board the M2-F2 lifting body while wearing a standard flight suit and helmet. The mission he was preparing to undertake was not planned as a high-altitude mission. (NASA)

---

4. R. Dale Reed and Darlene Lister, *Wingless Flight: The Lifting Body Story* (Washington, DC: NASA SP-4220, 1997), pp. 8–17.

and F-106. In 1962 he was the only civilian pilot selected to fly the X-20 Dyna-Soar spaceplane. When the Dyna-Soar project was canceled while still in the mockup stage, Thompson joined the X-15 program, eventually piloting 14 flights in the hypersonic rocket plane. In contrast to his high-speed (Mach 5.15) and high-altitude (214,100 feet) experience in the X-15, he also flew the Paraglider Research Vehicle, or Paresev, a diminutive craft that resembled a tricycle attached to a hang glider.[5]

First of the Heavyweights

Thompson was assigned as checkout pilot for the full-scale lifting body glider, designated M2-F1. Ground- and air-tow testing of this craft proved that a pilot could control the wingless configuration and maneuver it to a safe landing.

In the wake of initial success with the plywood M2-F1, work pressed forward on the M2-F2, first of the heavyweight, all-metal lifting bodies, which were more representative of the reentry design that would be necessary for returning space flights. The new shape was generally similar to that of the M2-F1, but the cockpit had to be moved forward of the center of gravity in order to accommodate fuel tanks to supply a four-chamber XLR11 rocket engine. In order to reach a suitable launch altitude, the M2-F2 was designed to be carried aloft beneath the wing of a modified B-52 Stratofortress. Following release, a typical flight profile included gliding maneuvers and landing or powered flight until engine burnout, and then a gliding touchdown on Rogers Dry Lake.[6]

Engineers were concerned when early M2-F2 design data predicted relatively low lateral stability, even though the aircraft was equipped with a stability augmentation system. One attempt to mitigate this problem involved the installation of an adjustable interconnect linkage between the rudders and the ailerons, which allowed the pilot to yield a variable amount of rudder or aileron movement. The pilot could adjust the control gains as needed during flight, using a special wheel installed on the left side of the cockpit.

For the initial flight, primary objectives included verification of the vehicle's basic handling qualities and safe landing on the lakebed runway. The flight plan called for release at 45,000 feet and an indicated airspeed of 165 knots. The pilot was to accelerate to 220 knots in a dive before executing a turn over the north end of the lakebed and increasing speed to 300 knots. He would then execute a simulated landing flare at 22,000 feet to verify that the lifting

---

5. Donald L. Mallick and Peter W. Merlin, *The Smell of Kerosene: A Test Pilot's Odyssey* (Washington, DC: NASA SP-4108, 2003) p. 101.

6. Reed and Lister, *Wingless Flight: The Lifting Body Story.*

body had sufficient lift to arrest the descent and satisfactory controllability at landing speeds.

The practice flare also gave the pilot an opportunity to evaluate predicted lateral-directional pilot-induced oscillation (PIO) problems that had been observed during simulations. Engineers expected the PIO problem to occur during high-angle-of-attack, high-speed flight such as that experienced during maneuvers for touchdown. The practice flare provided an opportunity to evaluate PIO tendency while still at an altitude high enough to permit corrective action. If the severity of the PIO exceeded expectations, the pilot could reset the aileron-rudder interconnect prior to the final landing flare, when there was little room for error.[7]

As dawn broke on July 12, 1966, technicians completed a flurry of final preparations for the maiden flight in the cooler predawn air. Milt Thompson boarded the M2-F2 wearing a standard flight suit and helmet, as this mission had not been planned to involve high altitudes. Pressure suits are not required for flights below 50,000 feet. Technicians assisted with final preparations for pilot life support, communication channels, and cockpit configuration.

The B-52 took off with the M2-F2 attached to a pylon on the underside of its right wing. This same pylon had been used to carry the X-15 rocket plane. An adapter allowed carriage of the lifting body.

All parameters were normal as NASA's Fitzhugh "Fitz" Fulton and Air Force copilot Jerry Bowline flew the B-52 mother ship to drop-altitude over Rogers Dry Lake. While the B-52 circled the lakebed, launch panel operator Vic Horton checked the lifting body's systems and Thompson completed his pre-flight checklist. First, Thompson activated the batteries that provided power to nearly every system in the

The M2-F2 was carried to launch altitude beneath the wing of a modified B-52. Pilot Milt Thompson endured the confines of the cramped cockpit with a smile. (NASA)

---

7. Milton O. Thompson and Curtis Peebles, *Flying Without Wings: NASA Lifting Bodies and the Birth of the Space Shuttle* (Washington, DC: Smithsonian Institution Press, 1999).

vehicle. Next, he activated hydraulic systems that drove the flight controls. He verified that the stability augmentation system was fully functional and checked various control-surface positions to ensure that each correlated with stick and rudder inputs. Having verified all systems were functioning properly, Thompson commenced the final countdown.[8]

"Okay. Five, four, three, two, one, release! Okay,"[9] Thompson said.

As the M2-F2 dropped away from the mother ship, Thompson experienced brief weightlessness. The vehicle rolled slightly, but he quickly leveled out and started the first maneuver. The turn progressed without incident, and the vehicle responded well in pitch and roll. As Thompson increased his airspeed to 300 knots in preparation for his simulated landing flare, he noticed a slight lateral-directional oscillation. In order to reduce PIO tendency, he adjusted the aileron-rudder interconnect ratio slightly, which seemed to alleviate the problem. During the practice flare, Thompson found it easy to maintain proper AOA and acceleration. What happened next, however, was entirely unexpected.

At 16,000 feet indicated altitude, Thompson rolled into a left turn for final approach to the runway. Sensing that he was close to experiencing a control-lability problem, he again adjusted the aileron-rudder interconnect wheel, setting the ratio to 0.4 as he began the second of two planned 90-degree turns. Roll response, however, was unexpectedly weak, so he attempted to increase the interconnect ratio to 0.6. To Thompson's surprise, the vehicle now developed a significant longitudinal roll oscillation just as he was lining up for final approach. Each deflection of the control stick resulted in an undesirably high roll rate. To compensate, he attempted to decrease the interconnect ratio, but the oscillations increased, rolling the vehicle about 45 degrees in either direction. Desperate to arrest the motions, Thompson moved the control wheel all the way to its stop, hoping to reduce the aileron-rudder interconnect ratio to zero. To his horror, the rolling motion increased violently, swaying from 90 degrees left to 90 degrees right in less than a second.[10]

The M2-F2 was in a 27- to 30-degree dive at 300 knots about 8,000 feet above the ground and descending at approximately 18,000 feet per minute. Less than 30 seconds from impact, the options available to the pilot were extremely limited. Thompson knew that he was experiencing PIO and that the quickest way to stop a PIO is (counter-intuitively) to stop making any control inputs and the oscillations should damp out themselves, but did he have time? Ignoring every natural instinct, he forced himself to let go of the

---

8. Ibid.

9. Ibid., p. 118.

10. Ibid.

control stick and then looked at the aileron-rudder interconnect wheel. He reasoned correctly that if moving the wheel in one direction had caused the vehicle to become overly sensitive to control inputs, then moving it in exactly the opposite direction should correct the problem. As soon as he did so, the oscillations abated. Having regained control, Thompson finished the flight by touching down smoothly on the dry lakebed.[11]

Habit Patterns and Negative Transfer of Training

In the postflight debriefing, Thompson admitted that he had unwittingly turned the interconnect wheel in the wrong direction. Apparently, when he had attempted to adjust the interconnect ratio down from 0.6, he had inadvertently increased it. Initial attempts to fight the PIO only aggravated the problem.[12]

But why did Thompson, an experienced research pilot who had trained extensively in the simulator prior to the flight, make such a rudimentary mistake? As it turned out, this was not simply an example of inadvertent error on the part of the pilot. The answer lay in the design of the simulator and demonstrates a worthwhile lesson about the importance of human factors in the design, testing, and evaluation of experimental aircraft.

As the M2-F2 approaches a runway on the dry lakebed at Edwards Air Force Base, a chase pilot observes from an F-104. During the maiden flight, the vehicle developed a significant longitudinal-roll oscillation just prior to final approach, but the pilot regained control in time to execute a safe landing. (NASA)

---

11. Ibid.

12. Reed and Lister, *Wingless Flight: The Lifting Body Story.*

To begin with, M2-F2 project engineers had decided to use an Air Force X-15 simulator and retool it for their purposes. To do so, they turned the X-15's speed brake handle into a control for the M2-F2's aileron-rudder interconnect ratio. In the actual research vehicle, however, the handle was replaced with a wheel to control the ratio.[13] Almost unbelievably, from today's modern design perspective, the direction of the handle movement in the simulator required to either increase or decrease the ratio was exactly opposite that of the wheel in the aircraft. In effect, Thompson had been training in a simulator with handle movement that was completely opposite of the control movement for the desired response as found in the actual aircraft. This training design defect was not the result of any individual's lack of oversight; at that time there was simply no review or oversight process to ensure that such human factors were appropriately addressed.

Yet for the test pilot, practice in a simulator sets up a reinforced cognitive habit pattern that is later transferred to his actions in the aircraft. In a very real sense, "muscle memory" is also formed and reinforced as part of the neuromuscular control system. Neuroscience and human factors research suggest that such memory-response pathways are part of long-term memory. Over time, the practiced motions become automatic and no longer require detailed, conscious processing of information to perform the action. Instead, higher executive cognitive functions simply set the practiced neuromuscular programs in motion.[14] To manipulate aircraft controls positioned differently from those used previously in the simulator, the pilot must employ more cognitive resources to undo or counter previously developed behaviors, troubleshoot, and then correct the problem. In his memoir, Thompson reflected on the flight:

> An aircraft's controls should always be reproduced accurately in the simulator cockpit. Many serious problems have been encountered when the two sets of controls didn't match. The pilot spends many hours in the simulator preparing for flight, especially the first flight. He learns to control the vehicle instinctively. He doesn't have to look at a control to know what it is or which way he should move

---

13. Ibid.

14. Daniel A. Cohen, Alvaro Pascual-Leme, Daniel Z. Press, and Edwin M. Robertson, "Off-Line Learning of Motor Skill Memory: A Double Dissociation of Goal and Movement," *Proceedings of the National Academy of Sciences* no. 102 (2005): 18237–18241; P. De Weerd, K. Reinke, L. Ryan, T. McIsaac, P. Perschler, D. Schnyer, T. Trouard, and A. Gmitro, "Cortical Mechanisms for Acquisition and Performance of Bimanual Motor Sequences," *Neuroimage* no. 4 (2003): 1405–1416; D.B. Willingham, M.J. Nissen, and P. Bullemer, "On the Development of Procedural Knowledge," *Journal of Experimental Psychology, Learning, Memory, and Cognition* no. 6 (1989): 1047–1060.

it to get the desired result. He adjusts various controls by feel rather than sight. If the simulator does not duplicate the airplane, the pilot has to stop and think about what control lever he is touching and which direction he must move it. That could mean a deadly delay or a deadly error, as it almost did in this case.[15]

Thompson's dramatic experience underscores the importance of the field that today is known as human factors engineering and the relevance of such thinking throughout the flight research process. Thompson's close call in the M2-F2 in 1966 holds lessons that are relevant to today's activities in aerospace engineering, design, and testing of flight vehicles. These lessons apply to both crewed and robotic vehicles.

Although the aviation community has increasingly come to appreciate the role of human factors in piloted aircraft, there is a danger that in the new environment of unmanned aerial vehicles (UAV), some engineers may fail to fully appreciate these factors' importance. For example, the design of a UAV ground-control station—particularly when operated by professional pilots—cannot be haphazard or left to chance. The placement of displays and controls, the operation of alarms and warnings, and the operation of automated flight-management systems all must bear some consistency with those aircraft systems produced over long years of human factors experience so as to not violate well-learned pilot background experience and known stereotypes. To do otherwise invites mishap.[16]

---

15. Thompson and Peebles, pp. 126–127.

16. Anthony P. Tvaryanas, William T. Thompson, and Stefan H. Constable, "Human Factors in Remotely Piloted Aircraft Operations: HFACS (Human Factors Analysis Classification System) Analysis of 221 Mishaps over 10 Years," *Aviation, Space and Environmental Medicine* no. 77 (2006): 724–732; Christopher D. Wickens, John D. Lee, Yili Liu, and Sallie E. Gordon Becker, *An Introduction to Human Factors Engineering*, 2nd ed. (Upper Saddle River, NJ: Pearson Prentice Hall, 2004), pp. 1–9, 184–242, 418–435; Tovey Kamine and Gregg A. Bendrick, "Visual Angles of Conventional and a Remotely Piloted Aircraft," *Aviation, Space, and Environmental Medicine* (2009).

The Space Shuttle orbiter Enterprise was used for NASA's Approach and Landing Test Program 4 years prior to the first orbital flight tests. On the first few flights, an aerodynamic fairing covered the vehicle's dummy rocket engines. (NASA)

CHAPTER 3

Pilot-Induced Oscillation During Space Shuttle Approach and Landing Tests

From the earliest days of piloted flight, aviators have experienced a frightening and sometimes deadly phenomenon that has come to be known as pilot-induced oscillations, or PIO. The first recorded incident took place in 1903 when Wilbur and Orville Wright experienced mild longitudinal oscillations during flight testing of their Wright Flyer, the world's first successful powered aircraft.[1]

A Department of Defense handbook on aircraft flying qualities defines PIO as sustained or uncontrollable oscillations resulting from the efforts of the pilot to control the aircraft.[2] It is a complex interaction between a pilot and his active involvement with an aircraft feedback system, specifically the flight controls where pilot command input yields an inadvertent, sustained oscillation response. It is a form of closed-loop, or negative-feedback, instability.[3] Because it frequently results from a mechanical or software fault embedded in the FCS, some aircraft-pilot systems have a built-in predisposition to PIO.[4]

Human factors engineers and control theory experts typically describe three components that are required to produce a PIO: input, time lag, and gain. Input refers to commands (or other influences) that are introduced into the control system to produce an expected result. In an aircraft, this would normally be what the pilot accomplishes by moving the rudders and the control

1. Holger Duda, *Effects of Rate Limiting Elements in Flight Control Systems—A New PIO Criterion* (Reston, VA: American Institute of Aeronautics and Astronautics [AIAA] 95-3204-CP, 1995), p. 288.
2. Department of Defense, *Flying Qualities of Piloted Aircraft* (Washington, DC: U.S. GPO MIL-HDBK-1797, 1997), p. 151.
3. Wickens, Lee, Liu, and Becker, *An Introduction to Human Factors Engineering*, p. 238.
4. David H. Klyde et al., "Unified Pilot-Induced Oscillation Theory," vol. 1, "PIO Analysis with Linear and Nonlinear Effective Vehicle Characteristics, Including Rate Limiting" (Wright-Patterson Air Force Base, OH: Air Force Research Laboratory WL-TR-96-3028, 1995), p. 14.

stick or yoke. Other inputs, such as thrust adjustments and crosswinds, may affect controlled flight of the vehicle. The frequency of input commands is expressed in terms of cycles per second in Hertz (Hz). It is generally difficult for humans to track tasks with an input rate of 1.0 Hz or higher; most systems require input at a rate of about 0.5 Hz.[5]

Control order is a key element of input. The control order of a system relates a change in the command input to either a change in position (zero-order), a change in velocity (first-order), or a change in acceleration (second-order). Thus, the control order associates the input with the rate of change of the output of the system.

More specifically, strictly second-order systems, in which acceleration is changed as a result of input, are usually difficult to control because they can be unstable. The operator must therefore anticipate and predict elements of the system (i.e., render input based on a future state, rather than a present state), and this requires a high cognitive workload on the part of the operator; otherwise, unstable control is usually the result.[6] Second-order systems have to be carefully designed. However, systems with high mass and inertia, especially when falling or gliding downward—such as the Space Shuttle Approach and Landing Test (ALT) program vehicle—typically have sluggish acceleration responses to position inputs and hence can be difficult to control precisely.

The second element of PIO is time lag, or the duration from the time of a given input to the point when the desired output is realized. In airplanes equipped with cables that move the aerodynamic control surfaces, the time lag between input and output would be close to zero. In modern aircraft, however, with computer-controlled electronic FCSes, the time lag may be of longer, or even variable, duration. This can be especially relevant when competing inputs are received by the FCC(s) simultaneously, and priority is given to one set of inputs at the expense of another. Unacceptable delays in the output may result. In any event, longer time lags require greater anticipation for effective, precise control. Such anticipation is a source of increased cognitive workload on the part of the operator until the system's responses are fully appreciated and well practiced.

The third element of PIO is gain. System gain refers to the degree of output rendered for a given amount of input. In aircraft systems, gain is the amount (amplitude) of deflection of a flight control surface for a given amount of input by the pilot. This gain may even change depending on the speed of the aircraft; at very high speeds, a given input would result in very little control-surface

---

5. Wickens et al., *An Introduction to Human Factors Engineering*, p. 234.

6. Ibid., p. 237.

deflection, whereas at very slow speeds, the opposite would be the case. Instability can result if the gain is too high for a given set of conditions and circumstances. It should be noted that gain is a concept entirely separate from that of control order. In fact, a given degree of gain can be applied to any order system, whether zero-order, first-order, or second-order.

Given these three elements, a type of closed-loop instability manifesting as PIO may result from a combination of several factors: (1) there is more than sufficient time lag between the controller's input and the system's response; (2) the gain is too high for the conditions or circumstances; and (3) there is high input rate relative to the time lag of the system, which may be a secondary effect of an incorrectly blended control order.[7]

Conversely, designers can offer four solutions based on the principles of human factors engineering to reduce or eliminate closed-loop instability. First, the system gain can be lowered. Second, the time lag of the system can be reduced. Third, the operator can be trained to deliberately not correct every input result (thereby filtering out high-frequency input). This principle relates to the well-known pilot adage that the best thing to do in a PIO is to "get off the stick." Indeed, PIO is often made worse because the pilot inputs are out of phase with good control practices, resulting in increasing amplitudes at each cycle of attempted control. Finally, the operator can render input based on the anticipation of a future state of the system, although this may be done more effectively by computer-controlled systems.[8]

The term "PIO" is often considered pejorative because it seems to lay blame primarily on the pilot when, in fact, many other factors may be involved. Alternatively known as aircraft-pilot coupling, pilot-in-the-loop oscillations, or pilot-augmented oscillations, "PIO" is nevertheless the preferred term within the test-pilot and handling-qualities community.[9] Clearly the pilot is an unwilling participant, but pilot behavior is the source factor that distinguishes severe PIO from most aircraft feedback-control design problems.[10]

The severity of PIO can range from benign to catastrophic. A mild pitch bobble at altitude might cause consternation, but severe pitching during final

7. Ibid., p. 239.

8. Ibid., p. 239.

9. Joel B. Witte, "An Investigation Relating Longitudinal Pilot-Induced Oscillation Tendency Rating to Describing Function Predictions for Rate-Limited Actuators," AFIT/GAE/ENY/04-M16, (Wright-Patterson Air Force Base, OH: Department of the Air Force, Air University, Air Force Institute of Technology, March 2004).

10. Duane T. McRuer, *Pilot-Induced Oscillations and Human Dynamic Behavior*, NASA Contractor Report CR-4683 (Hawthorne, CA: Systems Technology, Inc., 1995).

approach to landing could be fatal.[11] Since the introduction of highly aug-mented, fly-by-wire controls in the 1970s, the potential for PIO has, in fact, increased due to the evolving nature of computerized FCSes.[12] During the first flight of the YF-16A aircraft (forerunner of the F-16 Fighting Falcon), PIO occurred inadvertently while conducting a high-speed taxi test because the control gains were set too high. Due to near loss of control, the pilot had no choice but to take off in order to avoid a ground accident.

The PIO phenomenon can be divided into three categories. Category I is associated with linear pilot-vehicle system oscillations, usually described as a low-frequency consequence of excessive high-frequency lag in an aircraft's linear dynamics.[13] Category II involves quasi-linear pilot-vehicle system oscil-lations associated with control-surface rate or position limiting.[14] Category III oscillations are associated with nonlinear pilot-vehicle systems and result from abrupt shifts in either the effective controlled-element dynamics or in the pilot's behavioral dynamics.[15]

In 1977, NASA research pilots experienced Category II oscillations during flight testing of a prototype Space Shuttle vehicle. Lessons learned from the incident, which took place during the critical landing phase, led to significant design changes to improve low-speed handling qualities of the operational orbiters. These changes were applied prior to orbital flight testing.

Spaceplane Prototype

The Space Shuttle orbiter was the first spacecraft designed with the aerody-namic characteristics and in-atmosphere handling qualities of a conventional airplane. In order to evaluate the orbiter's aerodynamic FCSes and subsonic handling characteristics, a series of flight tests known as the ALT program were undertaken at Dryden in 1977.

The ALT program demonstrated the capability of the orbiter to safely approach and land under conditions simulating those planned for the final phases of an orbital flight. The Dryden/Edwards AFB test site was selected because it included an instrumented test range with an extensive safety buffer zone and a 44-square-mile dry lakebed capable of supporting the landing

11. Mark R. Anderson and Anthony B. Page, *Multivariable Analysis of Pilot-in-the-Loop Oscillations*, (Reston, VA: AIAA-95-3203-CP, 1995), p. 278.

12. Brad S. Liebst et al., "Nonlinear Pre-filter to Prevent Pilot-Induced Oscillations Due to Actuator Rate Limiting," *AIAA Journal of Guidance, Control, and Dynamics* 25, no. 4 (2002): 740–747.

13. Klyde et al., "Unified Pilot-Induced Oscillation Theory," p. 17.

14. McRuer, *Pilot-Induced Oscillations and Human Dynamic Behavior*, pp. 79–80.

15. Klyde et al., "Unified Pilot-Induced Oscillation Theory," p. 17.

Enterprise, seen here during its first flight without the aerodynamic tail cone, was launched from atop a modified Boeing 747. These tests provided an accurate simulation of conditions expected following the orbiter's return from space. (NASA)

weight of the orbiter. Lakebed runways were used for the first four landings, and the fifth ended on a 15,000-foot concrete runway.[16]

Rockwell International built a full-scale orbiter vehicle prototype named Enterprise for the ALT. With a length of 122 feet, a wingspan of 78 feet, and a weight of 150,000 pounds, it was comparable in size and weight to a commercial transport aircraft. The majority of the orbiter's structure was made of aluminum alloys. Since it would not be subjected to reentry heating, Enterprise was not covered with the Space Shuttle's reusable surface insulation. It was instead covered with other materials, primarily polyurethane foam and fiberglass, in order to maintain the mold lines for aerodynamic purposes. The flight deck consisted of two crew stations for the commander (left side) and pilot (right side), with displays and controls allowing a single crewmember, operating from either station, to land the vehicle.

Aerodynamic controls included a body flap at the aft end, wing elevons, and a split rudder that doubled as a speed brake. Reaction-control systems,

---

16. "Press Kit: Space Shuttle Orbiter Test Flight Series," release no. 77–16 (Washington, DC: NASA, February 4, 1977).

unnecessary at low altitude, were not installed. For the captive flights and the first three free flights, an aerodynamic fairing covered the orbiter's aft end. On the last two flights, three dummy main engines were installed to simulate weight and aerodynamic characteristics of the operational orbiter.[17]

The vehicle was designed with a planned landing speed of about 185 knots and could be landed manually or by computer. An autoland system allowed the computer to guide the orbiter to the runway by determining its heading and speed using the inertial navigation system, microwave scanning-beam landing system, and other data sources.

A Boeing 747 airliner was modified as a Shuttle Carrier Aircraft (SCA) to carry Enterprise to altitude for the captive and free flight tests. Most of the passenger accommodations were removed, and parts of the fuselage underwent structural reinforcement to support the orbiter's weight. Tip fins were added to the horizontal stabilizers. Support struts (two aft and one forward) were installed atop the 747's fuselage to hold the orbiter. At the beginning of each free flight, explosive bolts released the orbiter from the SCA.

Free Enterprise

NASA selected two two-man orbiter crews for the ALT. The first consisted of Fred W. Haise, Jr. (commander), and Charles Gordon Fullerton (pilot), the second of Joe H. Engle (commander) and Richard H. Truly (pilot).

Crewmembers for the 747 SCA included pilots Fitzhugh "Fitz" Fulton and Thomas C. McMurtry and flight engineers Victor W. Horton, Thomas E. Guidry, Jr., William R. Young, and Vincent A. Alvarez.[18]

The ALT program allowed researchers to conduct a complete operational check of the orbiter's systems and provided experience that could not be gained through wind tunnel tests or simulation. Most important, it gave the crews hands-on experience and familiarized the pilots with the cockpit systems and the "procedural aspects of landing under conditions that are much easier to control than on the Orbital Flight Tests."[19]

The ALT program consisted of a series of incremental steps leading up to a final free flight demonstrating the orbiter's capability to land on a paved runway under conditions similar to those anticipated at the end of an orbital mission.

The first phase of the program involved airworthiness and performance verification of the modified 747. The next step consisted of three taxi tests

---

17. Ibid.

18. Ibid.

19. Herman A. Rediess, "Assessment of ALT Tests with Tailcone On Vs. Off," memorandum to DFRC Shuttle Program Manager from Director of Research, NASA DFRC, March 17, 1976.

NASA selected two orbiter crews for the Approach and Landing Test Program. The first consisted of Fred W. Haise, Jr., commander (left), and C. Gordon Fullerton, pilot. (NASA)

on the main runway at Edwards AFB. The SCA test team found no areas of concern that would prevent proceeding with flight tests.[20]

Taxi tests were followed by two sets of captive flights. Five captive-inactive flights were flown with an inert, uncrewed orbiter to verify the airworthiness of the 747 as an orbiter transport vehicle and to establish an operational flight envelope for subsequent captive-active and launch phases of the ALT program.[21] In the captive-active phase, the orbiter was powered up and piloted while mated to the SCA. Three captive-active flights verified the planned separation profile as well as orbiter stability and performance in the mated configuration, with combined operation of the primary FCS, auxiliary power units, hydraulics, and structure. Data from the flights demonstrated that the operational separation profile and procedures were satisfactory for the first planned free flight.[22]

---

20. Approach and Landing Test Evaluation Team, "Space Shuttle Orbiter Approach and Landing Test—Final Evaluation Report," JSC-13864 (Houston, TX: NASA Johnson Space Center [JSC], February 1978).

21. William H. Andrews, "Space Shuttle Orbiter Approach & Landing Test—Mated Inert Flight Test Plan" (NASA DFRC, January 28, 1977).

22. "Space Shuttle Orbiter Approach and Landing Test—Final Evaluation Report," JSC-13864.

The final phase of the ALT program included five free flights in which the orbiter was released from the SCA and glided to a landing at Edwards AFB. These tests demonstrated the capability of the orbiter to safely approach and land on a runway in a variety of center-of-gravity configurations within the operational flight envelope. The first four flights ended on airstrips marked on the dry lakebed. The final flight concluded with touchdown on a concrete runway in order to obtain data on the tire-pavement interface and qualify the deceleration system.

The first free flight on August 12, 1977, was piloted by Haise and Fullerton and successfully demonstrated the orbiter's basic low-speed handling qualities during descent and landing. The crew performed steering, braking, and coasting tests during the 11,000-foot rollout.[23]

During the second free flight, Engle and Truly (the crews alternated for each flight), performed various programmed stick inputs for FCS and structural evaluations. During heavy, moderate, and differential brake application, a "chattering" phenomenon was experienced as the natural frequency of the gear struts resonated with the antiskid control gains. Modifications of the antiskid system eventually resolved this problem.

For the third flight, the orbiter's center of gravity was moved aft to simulate tail cone–off stability characteristics. As on the previous flight, the crew

The Space Shuttle orbiter Enterprise on a steep approach toward the runway at Edwards Air Force Base during the first tail cone–off flight of the Approach and Landing Test Program. (NASA)

23. Ibid.

accomplished windup turns and performed test inputs and aerodynamic stick inputs. Closed-loop automatic guidance was employed after the final turn before landing.

During the first tail cone–off flight, the orbiter crew performed an angle-of-attack sweep and aerodynamic stick inputs to collect data on performance, stability and control, and flight handling qualities. The stage was now set for the final test involving a landing on the concrete runway.[24]

Flightcrew Qualifications

Because of the alternating crew schedule, Haise and Fullerton were slated to pilot Free Flight 5 in October 1977. As mission commander, Haise would occupy the left seat—typically considered the pilot's seat—while Fullerton would occupy the traditional copilot position on the right.

Haise, a native of Biloxi, MS, received a bachelor of science degree with honors in aeronautical engineering from the University of Oklahoma in 1959. He later received an honorary doctorate of science from Western Michigan University in 1970 and attended Harvard Business School in 1972.

His flying career included active service in the Navy, Marine Corps, Air Force, and NASA. Haise began as a Naval Aviation Cadet at Naval Air Station Pensacola, FL, in October 1952. From March 1954 to September 1956, he served as a Marine Corps fighter pilot, flying the F2H-4 Banshee, at the Marine Corps Air Station Cherry Point, NC. He also served as tactics and all-weather flight instructor at the Navy's Advanced Training Command at Naval Auxiliary Air Station Kingsville, TX.[25]

In March 1957, Haise joined the Oklahoma National Guard as a fighter interceptor pilot. In September 1959, he became a research pilot at the NASA Lewis Research Center (now the Glenn Research Center) in Cleveland, OH, where he conducted research on the flying qualities of various general-aviation aircraft and the use of aircraft for zero-gravity experiments. During this time, Haise also served in the Ohio Air National Guard and was recalled to active duty as a fighter pilot and chief of standardization and evaluation from October 1961 to August 1964.

In March 1963, Haise transferred to the NASA Flight Research Center at Edwards AFB. There he flew a variety of fixed-wing general-aviation aircraft; the Bell Model 47G helicopter; jets including the T-33, T-37, and F-104; and such specialized research aircraft as the NT-33A Variable-Stability Trainer and

---

24. Ibid.

25. Fred W. Haise, Jr., biographical files, Dryden Historical Reference Collection, NASA DFRC, Edwards, CA.

the M2-F1 lifting body. In 1964, he completed postgraduate courses at the U.S. Air Force Aerospace Research Pilot School, receiving the Honts Trophy as outstanding graduate of his class.

In April 1966, Haise was one of the 19 astronauts selected by NASA for the Apollo program. He served as backup Lunar Module pilot for the Apollo 8 and 11 missions and backup spacecraft commander for the Apollo 16 mission.[26]

Haise was assigned as Lunar Module pilot on the ill-fated Apollo 13 mission that was scheduled for a 10-day expedition to the Moon. The original flight plan, however, had to be modified en route due to a catastrophic failure of the Service Module cryogenic oxygen system that occurred at approximately 55 hours into the flight. Haise and fellow crewmembers—James Lovell (spacecraft commander) and John Swigert (Command Module pilot)—working closely with ground controllers, converted their Lunar Module into an effective lifeboat. Their successful efforts to conserve both electrical power and water assured their safety and survival while in space and for the return to Earth.

A fellow of the American Astronautical Society and the Society of Experimental Test Pilots, Haise flew more than 80 types of aircraft and eventually accumulated 9,300 hours' flying time, including 6,200 hours in jets. He also logged 142 hours and 54 minutes in space. From April 1973 to January 1976, he served as technical assistant to the manager of NASA's Space Shuttle project. He was then selected to command three of the five ALT flights.[27]

Fullerton earned a master of science degree in mechanical engineering from the California Institute of Technology, Pasadena, CA, in 1958 while working as a mechanical design engineer in the Flight Test Department of Hughes Aircraft Co., Culver City, CA. He joined the Air Force in July 1958.

After flight school, Fullerton was trained as an F-86 interceptor pilot, and later as a B-47 bomber pilot. In 1964, he was selected to attend the Aerospace Research Pilot School at Edwards AFB. Upon graduation, he was assigned as a test pilot with the Bomber Operations Division at Wright-Patterson AFB, Dayton, OH.[28]

Fullerton was selected in 1966 as a flightcrew member for the Air Force Manned Orbiting Laboratory (MOL), a classified program to develop a crewed spaceborne reconnaissance platform. Following the cancellation of the MOL in 1969, Fullerton joined NASA. After assignment as an astronaut, he served

---

26. Ibid.
27. Ibid.
28. Charles Gordon Fullerton biographical files, Dryden Historical Reference Collection, NASA Dryden Flight Research Center, Edwards, CA.

on support crews for the Apollo 14, 15, 16, and 17 lunar missions. In 1977, Fullerton was assigned to one of the two ALT flightcrews as orbiter pilot.

Following the ALT project, Fullerton went on to pilot the third Space Shuttle orbital flight-test mission in March 1982—the only Shuttle flight ever to land at White Sands, NM. Northrup Strip, later renamed White Sands Space Harbor, was used as an alternate landing site because Rogers Dry Lake at Edwards AFB was wet due to heavy seasonal rains. Fullerton later commanded the Space Transportation System (STS)-51F mission, with the orbiter Challenger carrying the Spacelab module. The mission ended August 6, 1985, with a landing at Edwards AFB. During his two Shuttle missions, Fullerton logged 382 hours of space flight.[29]

PIO Incident

The fifth free flight of Enterprise took place on October 26, 1977, before a large crowd of news media, guests, and dignitaries that included Prince Charles, heir to the British throne. Throughout the ALT series, the orbiter had performed well mechanically and structurally, verifying preflight aerodynamic predictions. Perhaps most significantly, the final free flight revealed the orbiter's

Enterprise approaches the concrete runway at a speed of 200 knots. Pilot-induced oscillation developed during touchdown. (NASA)

---

29. Ibid.

susceptibility to PIO, prompting NASA engineers to develop a way to fix the problem before the first actual orbital flight test.[30]

Separation of the orbiter from the SCA occurred at an altitude of 17,000 feet and at 245 knots airspeed. Haise and Fullerton guided the vehicle on a straight-in approach to the 15,000-foot concrete airstrip, controlling the entire approach and landing sequence manually.[31]

As Enterprise approached the Edwards AFB main runway, Fullerton adjusted the vehicle's attitude in order to increase airspeed to 290 knots. Meanwhile, Haise made left rudder and roll inputs prior to speed-brake deployment to complete a set of aerodynamic data requirements. Enterprise intercepted the glide slope at 9,600 feet above ground level while the crew maintained alignment with the surface aim point and correlated instrument readings. In a smooth transition, as the orbiter dropped through 7,000 feet, Haise assumed control of the speed brake. When Enterprise was 4,000 feet above ground level, the crew noticed that they had drifted above the glide slope. To reacquire the aim point and prevent overspeed, Haise pitched the nose over and deployed speed brakes to 80 percent. He responded to a momentary 10-knot airspeed decrease by reducing speed brakes slightly. Fullerton then noted an airspeed decrease to 275 knots followed by a rapid increase to 290 knots.[32]

At 2,000 feet above the ground, the orbiter was on a slightly steep trajectory but still aligned with the aim point. Indicated airspeed was 294 knots (4 knots higher than planned), and the orbiter was 600 feet closer to the runway threshold than planned. Haise delayed speed brake retraction to compensate for the excess speed, but there was a 7-knot tailwind. As the crew lowered the landing gear, Enterprise approached the runway threshold at the correct altitude but 20 knots faster than planned. Haise set the speed brakes to 50 percent, anticipating that the vehicle would be slow at touchdown, but it continued to remain high at 200 knots as it approached within 500 feet of the touchdown line.

The orbiter seemed to float for an uncomfortably long time at an altitude of 4 feet above the runway. Haise attempted to overcome this with forward

---

30. Peter W. Merlin, "Free Enterprise: Contributions of the Approach and Landing Test (ALT) Program to the Development of the Space Shuttle Orbiter," AIAA-2006-7467, presented at American Institute of Aeronautics and Astronautics Space Conference, San Jose, CA (September 21, 2006).

31. Ibid. Note: Each crew station is equipped with identical rotational hand controllers. In pitch, the controller pivots about the palm of the hand; in roll, it pivots about a point slightly below the base. The two controllers are not mechanically linked, but if both are deflected, the input signals are combined and the elevons respond accordingly.

32. Ibid.

rotational hand-controller commands, but to no effect. The orbiter's altitude then ballooned slightly before the vehicle touched down smoothly some 1,000 feet beyond the planned point and at a speed of 180 knots. Enterprise skipped upward into the air and rolled to the right. As Haise attempted to level the wings, a lateral PIO developed. As the oscillations continued, the crew realized that the roll commands through the rotational hand controller were abnormally large and control response was lagging. Haise discontinued command inputs and allowed the roll rate to damp out to nearly wings-level. About 6 seconds after skipping into the air, thanks to timely recognition of the situation and compensation by the crew, the Enterprise touched down for a second time and rolled safely to a stop.

Analyzing the Problem

Data from Free Flight 5 indicated that the PIO resulted when Haise's inputs to control sink rate caused an unexpected pitch oscillation during the last 8 seconds prior to initial touchdown. The control system software limited the elevon rate to 26 degrees per second in order to cope with hydraulic flow limits for the elevon actuators. Moreover, because the system was designed to give priority to pitch inputs, the FCS failed to respond quickly to some roll inputs. This resulted in 4 seconds of PIO as the orbiter touched down gently with wings level and then skipped back into the air while rolling to the right. Haise ceased roll input momentarily, allowing the motion to damp out immediately prior to a second touchdown 6 seconds after the first. The left wheel bounced momentarily, but the vehicle quickly settled into a normal rollout. After nose-wheel touchdown, the crew applied light braking until the orbiter decelerated to 100 knots, heavy braking down to 50 knots, and finally light braking again until the vehicle came to a stop. The first main gear touchdown occurred 1,000 feet beyond the planned touchdown point, and the final touchdown was 1,900 feet farther. Total runway rollout distance from the initial touchdown was 7,930 feet.[33] The crew was forced to accept a higher-than-normal sink rate because of concern about airspeed bleed-off to 155 knots. Consequently, the orbiter landed harder than planned. The left main wheel also lifted slightly on the rebound but quickly settled onto the runway. The crew fully opened the speed brakes and applied differential braking to bring the orbiter to a stop.[34]

Pilot inputs to control sink rate resulted in large elevon motion (12 degrees peak-to-peak) at 0.6 Hertz and kept the elevons rate-limited throughout most

---

33. "Space Shuttle Orbiter Approach and Landing Test—Final Evaluation Report," JSC-13864.

34. Ibid.

of the touchdown sequence. This overrotation caused the orbiter to skip back into the air. The pilot was unaware of any problem beyond the fact that he was landing long. He applied a forward stick input to halt the ballooning and inadvertently initiated a roll command, possibly due to the unusual control stick geometry (stick longitudinal axis is inclined to the vehicle's axis). Because the center of pitch motion is near the cockpit, normal acceleration cues were lacking during small pitch oscillations. Also, due to cockpit visibility limitations, small changes in pitch attitude were not readily apparent to the crew. Consequently, neither crewmember detected the oscillation that caused elevon rate limiting.[35]

Haise attempted to correct the roll motion and applied forward stick to force the orbiter back down to the runway. This combination of control inputs saturated the control system, allowing phase lag to build up as the pilot continued to overcontrol in both pitch (to reestablish flightpath control) and roll (as a result of PIO). Because of the rate-saturated pitch channel, the FCSes priority rate-limiting design did not allow response to some roll inputs. The hydraulic system priority logic locked out any roll commands because pitch had priority. This triggered the large roll delay at touchdown and subsequent PIO. By releasing the controller momentarily, the pilot allowed the motions to damp out naturally just prior to the second touchdown.[36]

Of note is the fact that, just over 2 years prior to this event, NASA research pilot John A. Manke had landed the X-24B lifting body precisely on the white target landing spot at the 5,000-foot mark of the Edwards AFB runway. Afterward, he stated: "We now know that concrete runway landings are operationally feasible and that touchdown accuracies of plus or minus 500 feet can be expected."[37] Manke had demonstrated the feasibility of precisely landing an unpowered space reentry vehicle on a runway, thus creating high expectations for future Space Shuttle landing performance. Whether consciously realized or not, this accomplishment and his statement may have been a factor in the high rate of control-stick input rendered by the Enterprise commander in the final 8 seconds of the final ALT flight.[38]

---

35. Ibid.

36. Milton O. Thompson, notes on "Sequence of events on FF-5 landing," n.d., in the NASA DFRC Historical Reference Collection, File Folder L1-5-2-2, Milton O. Thompson Collection.

37. R. Dale Reed with Darlene Lister, *Wingless Flight: The Lifting Body Story* (Washington, DC: NASA SP-4220, 1997).

38. C. Gordon Fullerton, personal conversation with Dr. Gregg Bendrick at NASA Dryden Flight Research Center, Edwards, CA, June 5, 2006.

Operational Fix

Engineers at Dryden launched an all-out effort to comprehend and resolve the orbiter's PIO problem. In early 1978, Thompson, director of Dryden's Research Projects branch, drafted a plan to "obtain a current data base that will sharpen our awareness of all factors (subtle and obvious) that might influence a low L/D [lift-to-drag] orbiter runway landing in demanding situations." Part of the program called for the application of ALT data to computerized simulators for the purpose of familiarizing Shuttle pilots with the proper gain settings for landing. At the same time, Dryden's F-8 Digital Fly-By-Wire test bed—an ex-Navy fighter jet with highly modified flight controls—provided flight-test data to determine how delayed computer response to human input might be reduced or eliminated.[39]

Five research pilots flew PIO data flights resulting in 60 landings simulating the orbiter's control characteristics. They found that lags as short as 200 milliseconds between pilot input and discernible control-surface response profoundly impacted the aircraft's handling qualities. At the PIO condition, rate limiting decreased system gain and introduced phase lag into the system.[40]

To solve the problem, engineers developed a software filter that dampened the types of pilot inputs most likely to cause oscillations without affecting handling qualities or creating control time delays. These software changes worked, reducing PIO tendencies. Greater landing control, however, came at the expense of some degree of control-stick responsiveness.[41]

Additional studies were carried out using the Air Force's Calspan Total In-Flight Simulator (TIFS), a highly modified C-131H transport aircraft. The results characterized deficiencies in the orbiter's low-altitude longitudinal handling qualities contributing to the pilot's inability to precisely control flightpath angle and altitude change rate, predict aircraft response to control inputs, and adequately control the vehicle in disturbances due to external forces such as wind gusts.[42]

The TIFS flights allowed researchers to replicate the ALT flight control problem and develop control system modifications for incorporation into the

---

39. Michael H. Gorn, *Expanding the Envelope: Flight Research at NACA and NASA* (Lexington, KY: The University Press of Kentucky, 2001), p. 351.

40. John W. Smith, "Analysis of a Longitudinal Pilot-Induced Oscillation Experienced on the Approach and Landing Test of the Space Shuttle" (Washington, DC: NASA TM-81366, December 1981).

41. Gorn, *Expanding the Envelope*, pp. 352–353.

42. Robert G. Hoey et al., "Flight Test Results from the Entry and Landing of the Space Shuttle Orbiter for the First Twelve Orbital Flights" (Edwards, CA: Office of Advanced Manned Vehicles, AFFTC-TR-85-11, June 1985).

orbiter Columbia prior to the first orbital flight test.[43] Eight NASA research pilots (including five Shuttle astronauts) flew 16 2-hour flights in the TIFS. They completed 155 approaches (78 actual and 77 simulated) under flight conditions that included both actual and simulated air turbulence.[44]

Engineers determined that the orbiter had two modes affecting longitudinal control. The first involved the effective time delay between pilot input and vehicle response in pitch attitude control. As on most aircraft, the mechanical control actuators contributed a significant delay, as did the structural and smoothing filters required because of the high-gain feedback control system. The digital FCS also contributed delays because of the average sampling and computation time. The nonlinear control stick gearing provided good sensitivity around the neutral stick position while retaining a good maximum pitch rate or normal acceleration capability, but it contributed to the orbiter's pitch attitude PIO tendencies.

The second mode was altitude or flightpath control. Loss of lift caused by elevon deflection caused a nose-up pitch command to result in a downward acceleration at the center of gravity. Because the cockpit was located near the center of rotation, there was a 0.5-second delay before the pilot detected motion. The sluggish rise time of acceleration to its steady-state value, combined with delayed perception of motion, made it difficult for the pilot to accurately control attitude. High cockpit location and poor forward visibility also contributed to the pilot's inability to judge both attitude and altitude near touchdown.[45]

A model-following accuracy and test technique developed during the TIFS simulations provided useful data for assessment of pilot performance in potentially off-nominal situations. Only 15 percent of the approaches received satisfactory handling qualities ratings from the pilots. Of the rest, 70 percent were rated unsatisfactory and the remaining 15 percent were considered unacceptable due to the orbiter's unforgiving longitudinal control characteristics, particularly in the landing phase. Pilots with minimal or no prior experience with orbiter flight simulations had severe difficulty achieving successful landings due to

---

43. C.R. Chalk and P.A. Reynolds, "Test Plan—Simulation of Orbiter Landing Characteristics in the USAF Total In-Flight Simulator (TIFS)," Calspan Corporation, TIFS Memo No. 844, May 25, 1978.

44. "TIFS Program Summary," n.d., in the NASA DFRC Historical Reference Collection, File Folder L1-5-2-2, Milton O. Thompson Collection. Also see Bruce G. Powers, "Space Shuttle Pilot-Induced-Oscillation Research Testing" (Washington, DC: NASA TM-86034, February 1984).

45. Bruce G. Powers, "Low-Speed Longitudinal Orbiter Flying Qualities," Space Shuttle Technical Conference, NASA CP-2342, Part 1, Johnson Space Center, Houston, TX (1983), pp. 143–150.

inability to perceive deviations and make desired corrections to the flightpath quickly enough. With extreme concentration, more experienced pilots found it easier to perceive deviations and avoid the need for large flightpath corrections. Additionally, they were able to develop a pulsing control technique to minimize rate-limiting problems.[46]

The orbiter's PIO tendencies were found to be considerably more noticeable in flight tests than in ground simulations. Based on the level of their experience, the TIFS evaluation pilots decided they desired a well-damped, but more responsive, airplane. This goal was achieved to some degree by increasing the pitch forward-loop gain and allowing unlimited elevon surface rates. Some evaluation pilots noted improved handling qualities when the cockpit was moved (via simulation) 40 feet forward of the center of pitch-rotation. Additionally, the conventional center stick provided improved control compared to the rotational hand controller.[47]

Lessons Learned and Orbital Flight-Test Results

Development of the Space Shuttle orbiter produced the first reusable spacecraft capable of returning from orbit and landing on a conventional runway. This pioneering effort forced engineers to confront complex challenges in developing a vehicle with longitudinal flying qualities required for landing the orbiter manually in an operational environment. The ALT program was the final hurdle before the first orbital mission.

Based on ALT flight data and orbiter crew evaluations, all objectives of the program were successfully accomplished. The PIO incident during Free Flight 5 led to follow-on research to correct the PIO problem. To improve chances of coping with deviations at landing (i.e., turbulence and crosswinds), the Approach and Landing Evaluation Team recommended the following:

1. The [orbiter's] energy state should be maintained at the preplanned nominal level throughout the flight trajectory using standardized pilot techniques or autoland.
2. The trajectory from preflare to touchdown should be optimized for manual control.
3. Operational and flight control system limits should be determined and verified by simulation to determine the crew and vehicle capabilities and limitations to perform a safe landing.

---

46. "DFRC Orbiter Landing Investigation Team Final Presentation," NASA DFRC Historical Reference Collection, File Folder L1-5-7-22, Milton O. Thompson Collection (August 1978).

47. Ibid.

4. The flight control system must be modified to always provide at least some combination of pitch and roll capability to allow manual and automatic control for landing.

5. Sensitivity of the flight control system to PIO should be reduced.

6. Nominal trajectory planning should not require the use of speed brakes after flare (due to increased crew workload in the flare).[48]

Using data obtained from fixed-base and in-flight simulations, NASA engineers developed reasonably effective PIO suppression filters for use on Columbia. Because the software revisions merely mitigated, but did not completely eliminate, the orbiter's latent PIO tendencies, NASA scientists continued to study the problem well into the 1980s, long after the Space Shuttle's first orbital test flights.

The orbiter Columbia launched into space for the first time on April 12, 1981. After completing a 2-day orbital checkout, the crew (John Young and Robert

Lessons learned during the Approach and Landing Tests culminated in the safe landing of the orbiter Columbia following its maiden orbital flight in April 1981. (NASA)

---

48. "Space Shuttle Orbiter Approach and Landing Test—Final Evaluation Report," JSC-13864.

Crippen) made a successful landing on the dry lakebed at Edwards AFB. In March 1982, the third flight, crewed by Jack Lousma and C. Gordon Fullerton, ended with a landing on a dry lakebed at the U.S. Army White Sands Missile Range, NM. During final approach, the autoland system was used all the way through the landing flare. Fullerton then took over for manual landing and experienced a hard touchdown because he perceived and reacted to a "heave mode" (positive pitch change) that had not, in fact, occurred.[49]

Analysis of flight-test results from the first 12 orbital Shuttle missions by the Office of Advanced Manned Vehicles at Edwards AFB concluded: "Presently, the Orbiter's insidious subsonic longitudinal handling qualities are considered acceptable considering the scope of the current STS program which relies heavily upon the skills of a relatively small and specially trained crew of astronauts." Air Force analysts felt the orbiter's handling difficulties were the result of vehicle configuration design rather than control system deficiencies. They recommended that future spacecraft designers conduct simulator investigations on the effects of cockpit location with respect to the longitudinal center of rotation of the vehicle.[50]

Although the PIO suppression filter (added prior to the first orbital mission) virtually eliminated high-frequency PIO tendencies, it was not designed to improve low-frequency, large-amplitude heave mode characteristics near touchdown produced by poor flightpath control. Nevertheless, orbiter crews had no significant PIO problems during the first 12 Shuttle landings. Extensive simulator training for orbiter pilots prevented all but a few isolated incidents of over-control tendencies in the subsonic longitudinal axis during the orbital flight tests and subsequent operational missions.

---

49. Hoey et al., "Flight Test Results"; comments of C. Gordon Fullerton to the author, August 2006.

50. Hoey et al., "Flight Test Results."

Part 2:
Physiological Factors

The rocket-powered X-15 was designed to explore characteristics of flight at the edge of space and at hypersonic speeds. Three vehicles flew 199 flights over the span of nearly a decade with only one fatal accident. (USAF)

CHAPTER 4

Screening Versus Design
The X-15 Reentry Mishap

On November 15, 1967, an Air Force test pilot lost his life while flying the rocket-powered X-15 research vehicle in a parabolic space flight profile. The pilot was known to have experienced vertigo (that is, spatial disorientation) while flying the X-15 on previous missions, and investigators initially focused on this factor as a potential cause of the mishap. Painstaking analysis revealed, however, that spatial disorientation was merely a contributing factor along with loss of mode awareness due to misinterpretation of a dual-use flight instrument. The resulting confusion between yaw and roll indications led to inappropriate flight control input and subsequent loss of aircraft control. A new analysis of the X-15 accident provides an understanding of the potential for spatial disorientation—particularly the oculoagravic illusion—associated with parabolic space flight profiles, as well as the importance of maintaining mode awareness in the context of automated cockpit design.

Research at the Edge of Space

The X-15 flight research program was a joint endeavor by NASA, the U.S. Navy, and the U.S. Air Force to develop and operate a crewed research vehicle capable of attaining speeds exceeding Mach 6.0 and altitudes above 60 miles. Initiated during the 1950s as a continuation of high-speed aerodynamic and thermodynamic research conducted by NASA's predecessor organization, the NACA, and the military services, the X-15 program produced a rugged, dependable research tool for probing the edge of space. For this bold endeavor, new control theory had to be created and tested in the very important, difficult, and critical transition zone between very high atmospheric altitude and space flight above the sensible atmosphere. Some aerodynamic pressure exists in this region, but not enough for complete aerodynamic control, so other methods such as augmentation with hydrogen-peroxide thrusters must be blended gradually into the control system as the atmospheric characteristics change.

Prime contractor North American Aviation of Culver City, CA, built three rocket-propelled X-15 research airplanes that eventually flew a total of

199 missions over a span of 10 years. Researchers used vast amounts of data from these flying laboratories in the development of several air and space vehicles, including the Space Shuttle. After meeting planned speed and altitude goals, researchers increasingly used the X-15 as a platform for additional scientific experiments to support civil and military programs.

North American test pilot A. Scott Crossfield made the first glide flight on June 8, 1959. The first powered flight followed on September 17, and the X-15 quickly exceeded all previous speed and altitude records. Capt. Robert White became the first man to fly Mach 4 on March 7, 1961, and the first to fly Mach 5 on June 23. He achieved Mach 6 5 months later and also became the first person to fly above 200,000 feet. White set a world altitude record of 314,750 feet on July 17, 1962. The following year, NASA pilot Joseph A. Walker set an unofficial record by reaching 354,200 feet (69 miles).[1]

Each X-15 flight profile was designed to accomplish either a high-speed or a high-altitude research objective. Typically, for a high-altitude profile, the X-15 would be carried to an altitude of 45,000 feet beneath the wing of a B-52. Following release, the pilot ignited the rocket engine and set a climb angle designed to achieve a specific altitude. The engine operated for about 85 seconds, propelling the X-15 to an altitude of around 185,000 feet. Following engine burnout, the vehicle would continue to coast upward to around 250,000 feet before succumbing to the downward pull of Earth's gravity. At peak altitude, more than 50 miles above Earth's surface, the X-15 was effectively in space—an environment where aerodynamic control surfaces ceased to exert any authority on the vehicle's movement. Above the sensible atmosphere, small hydrogen-peroxide-fueled thrusters in the X-15's reaction control system (RCS) provided attitude control. As the vehicle reached the peak of its ballistic arc and expended its kinetic energy, it began reentry into the atmosphere. At approximately 80,000 feet, the pilot would set up a steep approach trajectory toward a runway marked on a dry lakebed at Edwards AFB. The vehicle now became a very heavy glider with a low L/D. In the dense lower atmosphere, the pilot used aerodynamic control surfaces and energy management techniques to guide the craft to an unpowered landing.[2]

The three X-15 aircraft met a variety of fates. The third airframe built made its first flight on December 20, 1961, with Neil A. Armstrong at the controls. In the same aircraft in July 1962, Maj. Robert M. White set a world altitude record of 314,750 feet at a speed of Mach 5.45, earning his astronaut wings in the process. Seven other pilots also earned their astronaut wings in the third

---

1. Dennis R. Jenkins, *X-15: Extending the Frontiers of Flight* (Washington, DC: NASA SP-2007-562, 2007).

2. Ibid.

Typical X-15 Research Flight Paths

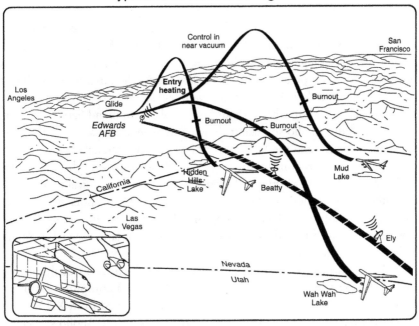

The three basic flight profiles for the X-15 were designed to investigate speed, altitude, and aerodynamic heating. A special radar tracking network, called the High Range, extended from California to Utah. (NASA)

X-15, including NASA research pilot Joseph A. Walker, who set an unofficial altitude record of 354,200 feet. The airplane was unfortunately destroyed several years later in the only fatal mishap of the program.[3]

The second X-15 aircraft, damaged in a November 1962 emergency landing, was rebuilt with improvements that allowed for greater fuel capacity and higher speeds. Redesignated X-15A-2, it was fitted with external fuel tanks and coated with an ablative heat shield. On October 3, 1967, Maj. William J. "Pete" Knight set an unofficial world speed record of 4,520 miles per hour (mph) (Mach 6.7). After he landed, it was discovered that the X-15A-2 skin and structure had suffered serious damage. It was subsequently retired to the U.S. Air Force Museum at Wright-Patterson AFB, OH.

The last eight flights of the program were made in the number one X-15. NASA pilot William Dana flew the final mission on October 24, 1968. Shortly

---

3. Peter W. Merlin and Tony Moore, *X-Plane Crashes—Exploring Secret, Experimental, and Rocket Plane Crash Sites* (North Branch, MN: Specialty Press, 2008).

afterward, the airplane was placed in the Smithsonian Institution's National Air and Space Museum.[4]

Numerous contributions to the design of modern air- and spacecraft, including that of NASA's Space Shuttles, were made through the use of aeronautical science and technological data gathered in the X-15 program. Less well known, however, were the program's contributions to the Apollo lunar missions. North American Aviation (later Rockwell International) served as the prime contractor for construction of both the X-15 and the Apollo Command Service Module. Designers of the Apollo spacecraft drew upon experience gained in the X-15 program and even used the X-15 as a test bed for new materials.

Materials and structures technologies developed for use in the X-15, especially those using titanium and Inconel X alloys, were applicable to Apollo and later spacecraft design. The discovery of localized "hot spots" on the X-15 led to the development of a bimetallic "floating retainer" concept to dissipate stresses in the X-15's windshield. This technology was later applied to Apollo and Space Shuttle orbiter windshield designs.[5]

X-15 reentry experience and heat transfer data were also valuable. Using their X-15 experience, Rockwell engineers designed a computerized mathematical model for aerodynamic heating. Lessons learned from X-15 turbulent heat transfer studies contributed to the design of the Apollo spacecraft; the discovery that the vehicles needed less thermal protection than had been thought led to lighter-weight future designs.

In 1967, technicians applied samples of cryogenic insulation to the X-15's speed brakes. Both adhesive and spray-on insulation, designed for use on the Apollo Saturn V rocket's second stage, were tested. The X-15 proved an excellent test bed for these materials because it could simulate the aerodynamic heating conditions that the Saturn rocket would face, and it allowed full recovery of equipment, calibration of results, and repeat testing where necessary.

Because the X-15 was a low-L/D vehicle, it presented a serious challenge with regard to approach and landing. Pilots and engineers worked together to develop intricate energy management techniques that eventually contributed to the development and operation of other low-L/D vehicles, such as lifting bodies and the Space Shuttle orbiters.[6]

---

4. Jay Miller, *The X-Planes: X-1 to X-45* (Hinckley, U.K.: Midland Publishing, 2001).

5. Peter W. Merlin, "Research Data from the X-15 Program Contributed to Apollo Lunar Missions," NASA Dryden Flight Research Center News Features, *http://www.nasa.gov/centers/dryden/Features/X-15_Apollo.html*, uploaded July 10, 2009, accessed January 15, 2010.

6. Ibid.

The X-15 lands on the dry lakebed at Edwards Air Force Base, with an F-104 chase plane close by. Smoke canisters provided wind direction data to the pilot. (USAF)

The X-15 program also produced a wealth of biomedical data that paved the way for human space travel. Researchers at the Air Force Flight Test Center Bioastronautics Branch and NASA's Flight Research Center (now Dryden Flight Research Center) made a careful study of X-15 pilots' heart and breathing rates to determine how they were affected at various critical points during flight. Despite initial concern, it soon became apparent that the pilots' higher heart rates were not associated with any physical problems or loss of ability to perform intricate mission-related tasks.[7]

A Stellar Career

Michael J. Adams was born in Sacramento, CA, on May 5, 1930. He enlisted in the U.S. Air Force in 1950 following graduation from Sacramento Junior College, where he was a varsity javelin thrower and baseball outfielder. After basic training at Lackland AFB, TX, he served with the 3501st Pilot Training as a Link Trainer instructor until being selected as an aviation cadet. Adams underwent primary training at Spence Field, GA, in October 1951. From

7. Ibid.

Maj. Michael J. Adams flew the X-15 seven times. During his final flight, he reached an altitude of 266,000 feet and qualified for astronaut wings, but he perished when the aircraft broke apart following reentry. (NASA)

Spence he went to Webb AFB, TX, for advanced training, where he earned his pilot wings and commission on October 25, 1952.[8]

Adams then transferred to Nellis AFB, NV, for gunnery school, where he flew the F-80 and F-86. Upon completion in April 1953, Adams was assigned to the 80th Fighter-Bomber Squadron at Suwon, South Korea. There he flew 49 combat missions and was awarded the Air Medal for meritorious achievements, accomplished with distinction above and beyond that expected of professional airmen, while participating in aerial flight operations.

After returning from Korea in February 1954, Adams spent 2 ½ years with the 813th Fighter-Bomber Squadron at England AFB, LA, plus 6 months' rotational duty at Chaumont Air Base in France.

Adams then entered the University of Oklahoma, Norman, OK, as part of an Air Force career development program for promising officers. He earned an aeronautical engineering degree in 1958 and later conducted graduate work in astronautics at the Massachusetts Institute of Technology, Cambridge, MA. After completing these studies, Adams went to work as an instructor for the Maintenance Officer Course at Chanute AFB, IL.[9]

During this time, he was selected as a student for the Experimental Test Pilot School at Edwards AFB. Adams graduated in 1962 and, as the outstanding pilot and scholar in his class, was awarded the Honts Trophy. He was then selected to attend the Aerospace Research Pilot School (ARPS), also at Edwards AFB, under the command of Col. Charles E. "Chuck" Yeager.

While attending the ARPS, Adams survived a landing accident in a two-seat F-104 by making a critical split-second decision. Adams was riding in the back seat of the airplane, which was piloted by fellow student Dave Scott,

8. Peter W. Merlin, "Michael Adams: Remembering a Fallen Hero," *X-Press* 46, no. 6 (July 30, 2004). NASA Dryden Flight Research Center, Edwards, CA (July 30, 2004): pp. 8–9.

9. Ibid.

who would later become a Gemini and Apollo astronaut and director of Dryden. As Scott was making a simulated X-15 approach for training and evaluation purposes, the F-104 suddenly lost power and began to drop rapidly. Both pilots realized that the jet would hit hard, and they made opposing decisions that saved their lives. Scott elected to stay with the airplane while Adams chose to eject. Adams pulled the ejection handle just as the F-104 slammed into the runway, breaking off its landing gear. His timing was perfect. Had he ejected before impact, his parachute would have had insufficient time to deploy due to the rapid rate of descent. If he had delayed ejecting for even a fraction of a second, he would have been crushed when the airplane's jet engine slammed forward into the rear cockpit. Adams's parachute opened just seconds before he hit the ground. He waved to Scott, who was climbing safely from the burning wreck. Scott's ejection seat had partially sequenced during the initial impact, locking his feet into the stirrups; had he ejected, he would have been killed.[10]

Adams graduated with honors from ARPS in 1963 and was subsequently assigned to conduct stability and control tests in the Northrop F-5A jet fighter. He later served as the Air Force project pilot on the Cornell Aeronautical Laboratory variable-stability T-33 program at Buffalo, NY. Adams was also one of four aerospace research pilots from Edwards AFB to participate in a 5-month series of NASA Moon-landing practice tests, beginning in January 1964, at the Martin Company in Baltimore, MD. The tests involved simulated lunar landing missions in a full-scale Command Module and a Lunar Excursion Module crew compartment mockup. Four simulated 7-day lunar landing missions were conducted, each with a three-man crew.

In October 1965, Adams was selected as an astronaut candidate for the Air Force Manned Orbiting Laboratory (MOL). MOL was designed to serve as an early space-based military reconnaissance platform. With the advent of the Corona uncrewed reconnaissance satellite program, Adams realized that there was little chance of an MOL flight in his future, and he requested a transfer to the X-15 program. In July 1966, he was accepted as the 12th and last pilot assigned to the program.[11]

During his first X-15 flight, on October 6, 1966, Adams achieved a speed of Mach 3, but a ruptured fuel tank caused premature engine shutdown 90 seconds after launch from beneath the wing of the B-52. Adams was forced to make an emergency landing at Cuddeback Dry Lake, about 40 miles northwest

---

10. Ibid.

11. Ibid.

of Edwards AFB. While on final approach to the lakebed, Adams remarked: "This thing is sort of fun to fly!"[12]

Ironically, Adams had a second emergency that day. During a routine proficiency flight in a T-38, he suffered an engine failure and had to make another emergency landing, this one at Edwards AFB.

Adams made his second X-15 flight on November 29, with only minor technical difficulties, and exceeded Mach 4, achieving a speed of 3,120 mph.[13]

During his third flight, on March 2, 1967, the aircraft lost cabin pressure while climbing through 77,000 feet, causing his pressure suit to inflate. Though this made it more difficult for Adams to fly the airplane, it was only the beginning of his troubles. As Adams arced the X-15 through a peak altitude of 133,000 feet and a maximum speed of Mach 5.59 (3,822 mph), his inertial computer failed, resulting in the loss of all velocity, altitude, and climb-rate readouts. Even without these data, Adams made a successful reentry and return to Edwards AFB. On approach, he radioed ground controller and fellow X-15 pilot Maj. William J. "Pete" Knight, saying: "I thought you said every *once* in a while something goes wrong, Pete."[14] In a postflight debriefing, Adams reported that he had suffered vertigo during the climb-out. This problem would return to haunt him again.

During his fourth and fifth flights, Adams encountered minor glitches but nothing unusual. On his sixth flight, the engine failed to ignite. Adams went through the restart procedure twice, finally igniting the rocket 16 seconds after launch. The rest of the flight was uneventful. His seventh and final flight ended in tragedy, cutting short a stellar career. At the time of his death, Adams had logged 4,574 flight hours. He had flown a wide variety of airplanes that included the T-6, F-80, F-84F, F-86, F-101, F-104, F-106, F-5, YAT-28, T-33, and T-38.[15]

Astronaut Wings Flight

Mike Adams piloted the 191st flight of the X-15 program on November 15, 1967. It was a mission designed to gather scientific data at high altitudes using the third X-15 vehicle, the same airplane in which 8 of the 12 X-15 project pilots had earned their astronaut qualifications by flying to altitudes in excess of 50 miles. Of the three vehicles, only this one—known as X-15-3—was equipped with an advanced control system as well as unique instrumentation

---

12. Radio transcript, X-15 flight report, Flight 1-69-116, October 6, 1966, NASA Dryden Historical Reference Collection, NASA DFRC, Edwards, CA.

13. Merlin, "Michael Adams: Remembering a Fallen Hero."

14. Ibid.

15. Ibid.

and displays specifically designed for high-altitude research. The Honeywell MH-96 adaptive FCS was run with a rudimentary computer that controlled the aircraft's direction and orientation. In the X-15-1 and X-15A-2 vehicles, the pilot normally controlled aerodynamic and ballistic flight phases manually using two different sets of flight controls. In the X-15-3, however, the MH-96 system eased pilot workload by automatically determining the optimal combination of aerodynamic and ballistic controls throughout the various phases of flight. When the X-15-3 was in atmospheric flight, the MH-96 moved the aerodynamic control surfaces in response to the pilot's input, and when flying above the sensible atmosphere, it activated the RCS to guide the airplane appropriately; the transition between the aerodynamic controls and the RCS thrusters was intended to be transparent to the pilot.[16]

For this flight the initial climb to altitude was uneventful. At 45,000 feet over Delamar Dry Lake, NV, the X-15 was released from the B-52 and Adams ignited the XLR99 rocket engine. He initiated a climb under full power but soon after passing 90,000 feet became aware of an electrical disturbance in the equipment that caused the aerodynamic flight control dampers to disengage. This disturbance also caused the MH-96 adaptive control system to disengage and, as a result, the automatic blending of the aerodynamic control surfaces and RCS thrusters ceased. Although Adams attempted to reengage the system, it continued to shut itself off. This meant he had to fly the airplane manually, as with the other two X-15s, applying aerodynamic controls and/or RCS thrusters using two different control sticks to maintain proper heading; this method of operation was manageable, but meant an unanticipated addition to the pilot's workload.[17]

As Adams continued his climb after engine shutdown, the test profile tasking (known as a "test card" to Air Force pilots) called for a slow wing rock of ten degrees above and below the horizontal in order for an experimental camera mounted on the X-15 to effectively scan the horizon. To assist the pilot with such maneuvers, the X-15-3 primary flight display had been modified so that when the appropriate switch was selected, the vertical bar on the display indicated precision roll rather than the normal yaw indication. The checklist called for return to the "alpha-beta" configuration (that is, yaw indication) at the conclusion of the wing-rocking (roll) maneuver. For reasons that have never been clear, the wing-rocking maneuver on this flight became excessive by a factor of two to three. Moreover, at the conclusion of the maneuver, the

---

16. Jenkins, *X-15: Extending the Frontiers of Flight.*

17. Donald R. Bellman et al., "Investigation of the Crash of the X-15-3 Aircraft on November 15, 1967," NASA Flight Research Center, Edwards, CA, 1968.

airplane began a slow drift in heading to the right, which was not corrected by Adams. After a brief time the X-15 was off by a heading of 15 degrees.[18]

Forty seconds later, and approximately 3 minutes after being launched from the B-52, the X-15 achieved its peak altitude of 266,000 feet. At this point, the drift in heading stopped and though the vehicle was moving forward, it was oriented in a yaw of 15 degrees to the right. The rightward drift in heading soon began again, and within 30 seconds the X-15 was flying in space at thousands of miles per hour with the nose of the aircraft pointed at a right angle to its flightpath (a 90-degree "beta"). Ground personnel in the flight control room, unfortunately, had no displays or any other way of discerning that the heading and alignment of the vehicle were off target; all such information was available only to the pilot. As a result, mission controllers were entirely unaware of the aircraft's true alignment and heading and therefore could not inform the pilot of the deviation.

Approximately 15 seconds later, Adams radioed that the aircraft "seems squirrelly." At approximately 230,000 feet and a speed of Mach 5, with the aircraft encountering increasing aerodynamic pressures as it descended into the atmosphere, Adams radioed that he was in a spin. This call was likely made after he saw a triangle of sunlight from the X-15 windscreen move across the cockpit in a horizontal fashion. This phenomenon clearly indicated a flat spin. Because flight control-room staff had no specific heading and alignment information, the only activities they could ascertain with their instruments were the aircraft's slow pitching and rolling motions. Adams radioed a second time that he was in a spin. There was no recommended hypersonic-spin recovery technique for the X-15, therefore little that control-room personnel could do to help rectify the situation.[19]

At 120,000 feet, however, the aircraft recovered from the spin and entered an inverted dive at Mach 4.7. This was likely due to the airframe's basic aerodynamic stability, though there may also have been some degree of pilot input. At this point, there was a chance for recovery, assuming Adams could bring the X-15 upright, roll out, and set up an emergency landing on a dry lakebed. But electrical transients now prevented the Honeywell adaptive flight control system from reducing the gains on the aerodynamic control surfaces as the vehicle entered the denser parts of the atmosphere. As a result, a full limit-cycle oscillation developed, causing the aircraft to engage in excessively severe pitch oscillations. The forces on the aircraft were 12 g's vertically and 8 g's laterally— possibly more—and it is unlikely Adams was conscious at this point. At about

---

18. Ibid.
19. Ibid.

Recovery crews study the wreckage of the X-15's forward fuselage. The vehicle exceeded structural design limits while traveling at a speed of Mach 3.93 at an altitude of 65,000 feet. The pilot did not eject. (NASA)

65,000 feet, with a speed of Mach 3.93, the aerodynamic forces on the X-15 exceeded structural design limits, and the vehicle broke apart.[20]

Adams was killed as a result of massive trauma due to impact forces. He had made no attempt to eject, probably because he was unconscious. In any case, he was far outside the operational envelope of the egress system, so chances of survival were virtually zero. Adams was the 27th American to fly more than 50 miles above Earth's surface and was awarded astronaut wings posthumously.[21]

NASA and Air Force officials convened an Accident Investigation Board, chaired by Donald R. Bellman of NASA. After 2 months of painstaking effort, the board concluded that Adams had allowed the aircraft to deviate in heading because of a combination of distraction due to pilot workload, misinterpretation of the instrument display, and possible vertigo.[22] A close review of these findings raises some questions about their veracity and/or completeness.

---

20. Ibid.

21. Jenkins, *X-15: Extending the Frontiers of Flight.*

22. Bellman et al., "Investigation of the Crash of the X-15-3."

A Question of Vertigo

The pilot's apparent lack of awareness regarding the aircraft's heading deviations, despite cockpit displays that were functioning normally, troubled investigators. The electrical disturbances in the early part of the flight certainly reduced the overall effectiveness of the aircraft's control system and thus increased the pilot's workload. Likewise, though the flight had had its share of technical glitches, this was not unusual for X-15 flights. Engineers had come to expect surprises during flights in the experimental craft because it occupied the cutting edge of aerospace technology at the time. As a potential explanation for Adams's erroneous control input, investigators began to focus on the question of whether or not vertigo may have been a contributing factor.

At the time of the X-15 program, the term "vertigo," or "pilot vertigo," was used to describe what contemporary medical practitioners now refer to as spatial disorientation associated with flight.[23] Unlike medical vertigo, pilot vertigo was not a description of disequilibrium experienced in normal activities as a result of any disease affecting the neurovestibular system.[24]

A background investigation revealed that Adams had, by his own admission, experienced pilot vertigo (that is, spatial disorientation) each time he flew the X-15. This was not abnormal because the acceleration thrust and climb angle during the boost phase of flight created conditions such that any pilot would experience vertigo as a normal physiological reaction. Fully fueled, the X-15 vehicle at ignition weighed approximately 33,000 pounds, whereas the XLR99 rocket engine was rated to generate approximately 57,000 pounds of thrust (though it frequently generated even more). This resulted in an initial axial acceleration of almost 2 g's, increasing to 4 g's at engine burnout.[25]

Moreover, since the X-15 was oriented in a markedly nose-up attitude during boost, the resultant vector of acceleration on the otolithic organs of the inner ear produced a somatogravic illusion.[26] Pilots more commonly describe this as "pitch-up illusion," "acceleration effect," or "takeoff illusion," a phenomenon in which the pilot feels as if the aircraft is pitching up even when the instruments clearly indicate a level trajectory. This effect was well known to

23. *Stedman's Medical Dictionary*, 27th ed. (Philadelphia, PA: Lippincott, Williams & Wilkins, 2000), p. 1711.

24. Richard G. Holt and Joe Ben Wiseman, "Otolaryngology in Aerospace Medicine," chapter 18 in *Fundamentals of Aerospace Medicine*, ed. Roy L. DeHart and Jeffrey R. Davis, 3rd ed. (Philadelphia: Lippincott Williams & Wilkins, 2002), pp. 420–427.

25. Bellman et al., "Investigation of the Crash of the X-15-3."

26. A.J. Parmet and Kent K. Gillingham, "Spatial Orientation," chapter 8 in *Fundamentals of Aerospace Medicine*, 3rd ed., pp. 220–221.

the X-15 pilots and usually lasted until the boost phase of the flight was over. Indeed, present-day naval aviators engaged in nighttime takeoffs from aircraft carriers in the F-18 launch hands-free in order to mitigate the desire to pitch the aircraft's nose down in response to the false pitch-up sensation.

Adams seemed to experience the sensation of what he termed "vertigo" for a much longer period of time than did other X-15 pilots. He had noted that he experienced vertigo throughout the duration of each flight up until reentry.[27] This meant he experienced such sensations after the axial acceleration had stopped. To better understand this particular effect, it is helpful to review some of the aeromedical research produced just a few years before Adams's mishap.

In 1952, Ashton Graybiel described a phenomenon associated with the somatogravic illusion, which he called the oculogravic illusion. Similar in nature to the somatogravic illusion, it also directly involved the eyes. Mediated by the vestibulo-ocular reflex, the oculogravic illusion resulted in an involuntary downward shifting of the eye gaze, with a corresponding upward shift in the visual field, in response to sustained linear acceleration. Pilots might describe this phenomenon as an inability to visually "hold" the instrument panel while attempting to look at it during periods of high acceleration. Thus, given the thrust and its duration, as well as the nose-up orientation of the X-15 during the boost phase of flight, this might be expected as an additional effect associated with the somatogravic illusion. However, even the oculogravic illusion should not have persisted long after the boost phase had stopped.[28]

Siegfried Gerathewohl and Herbert Stallings published a 1958 paper that seemed to address this. Utilizing data from parabolic flight profiles in high-performance aircraft, they demonstrated a reflexive upward shifting of the eye gaze—and a corresponding downward shift in the visual field—after the cessation of sustained axial acceleration, throughout the course of a parabolic flight profile. They termed this the "ocul*o*agravic illusion," referring to the effect on the eyes, mediated through the vestibular ocular reflex, due to the cessation of prolonged axial acceleration. Based on this evidence, it is therefore most likely that Adams experienced the oculoagravic illusion, which he described as vertigo lasting up to reentry.[29] According to modern theories regarding spatial disorientation, pilots are more susceptible to a variety of illusions when they

---

27. Milton O. Thompson, *At the Edge of Space: The X-15 Flight Program* (Washington, DC: Smithsonian Institution Press, 1992).

28. Ashton Graybiel, "Oculogravic Illusion," *Archives of Ophthalmology* 48, no. 5 (1952): 605–615.

29. S.J. Gerathewohl and H.D. Stallings, "Experiments During Weightlessness: A Study of the Oculoagravic Illusion," *Journal of Aviation Medicine* no. 29 (1958): 504–516.

are fatigued, task saturated, or distracted; when their attention is channelized; or when they are otherwise cognitively challenged, engaged, or compromised.[30]

As further support for the notion that Adams's neurovestibular system may have been particularly susceptible, there was objective evidence that he had a particularly sensitive labyrinthine apparatus of the inner ear. The results of certain medical screening tests during his astronaut selection physical at Brooks AFB, TX, were noted to have been "extremely abnormal."[31]

One particular test measured the duration of postrotatory nystagmus, otherwise known as a cupulogram, because it assessed the function of the cupula at the base of the semicircular canals in the inner ear. In this test, a candidate would be spun in a chair for several seconds at slow speed. The rotation would be stopped, and the duration of the involuntary reflex motions of the eyes, known as nystagmus, would be measured. Postrotatory nystagmus is a general indicator of the neurovestibular function of the inner ear, and the presumption was that if the duration of postrotatory nystagmus was too long, the pilot candidate had too great a sensitivity to angular accelerative forces on the body; hence, he would presumably be more likely to develop spatial disorientation. It is also possible that a more sensitive neurovestibular system under some circumstances allows for better detection of minor aircraft movements, leading to more proactive control (in theory). If Adams possessed this increased sensitivity, it might have been one of the reasons he was considered such a skilled pilot.

During and after World War I, the U.S. Army Air Corps (which later evolved into the U.S. Air Force) had a pilot requirement stipulating that the duration of postrotatory nystagmus be limited to between 16 and 36 seconds. However, during the early years of the space program, there was no established cutoff standard for cupulogram, and Adams was selected without prejudice as an astronaut for the Air Force MOL program.[32]

With regard to the X-15 accident, investigators were then left to ponder whether this vertigo, that is, the oculoagravic illusion, contributed to the mishap. Though not stated in such terms, this was apparently the assessment of the Accident Investigation Board's flight surgeon, Lt. Col. Robert Matejka, who subsequently recommended formal screening for vertigo (that is, labyrinthine sensitivity) in the pilot selection process. Presumably, this meant taking into consideration the duration of postrotatory nystagmus as an official screening criterion. In the accident board's report, Matejka writes:

---

30. Previc and Ercoline, "Spatial Disorientation in Aviation."

31. Bellman et al., "Investigation of the Crash of the X-15-3."

32. Maura Phillips Mackowski, *Testing the Limits: Aviation Medicine and the Origins of Manned Space Flight* (College Station, TX: Texas A&M University Press, 2006).

Tests for labyrinth sensitivity are not given routinely to X-15 pilots during their initial or periodic physical examinations. There are no established standards by which medical officers can rate the degree of susceptibility of a person to vertigo. Thus, the results of the tests on Major Adams [from Brooks AFB] were never placed on his medical records. In spite of the lack of such standards, it is believed that candidate pilots for the X-15 and comparable programs should be tested in this area and the results considered along with other factors that are not numerically definable.[33]

By today's standards, the term "vertigo" was imprecise in its reference to spatial disorientation, and its relationship to causality was even less well defined. Contemporary practitioners of aerospace medicine and physiologists refer to three specific types of spatial disorientation: Type I (Unrecognized), Type II (Recognized), and Type III (Incapacitating).[34] If the X-15 mishap was the direct result of spatial disorientation, it would have been almost certainly Type III (Incapacitating). Adams clearly knew he had experienced vertigo (a somatogravic illusion leading to mild spatial disorientation) whenever he flew the X-15, but he was not incapacitated by it as he had made a number of successful flights in the vehicle. Thus, this line of thinking suggests that he started out in Type II (Recognized) spatial disorientation, and as the rotational forces built up in this mishap, he probably transitioned to Type III (Incapacitating) spatial disorientation.

Historical evidence suggests, however, that Adams most likely experienced the oculoagravic illusion resulting from Type II (Recognized) spatial disorientation and did not transition to Type III (Incapacitating). Clearly, Adams knew he was prone to such sensations, and the cardinal rule of flight is that when a pilot recognizes spatial disorientation, he needs to "get on the instruments." Although the X-15 had windows through which the pilot could see the sky and/or horizon, it was not a vehicle that was flown using visual flight rules, so the pilot had to rely on instruments to control the vehicle.[35]

---

33. Bellman et al., "Investigation of the Crash of the X-15-3," p. 37.

34. A.J. Parmet and Kent K. Gillingham, "Spatial Orientation," *Fundamentals of Aerospace Medicine*; T.J. Lyons, W.R. Ercoline, J.E. Freeman, and K.K. Gillingham, "Classification Problems of U.S. Air Force Spatial Disorientation Accidents, 1989–91," *Aviation, Space, and Environmental Medicine* 65 (1994): 147–152.

35. Dennis R. Jenkins and Tony R. Landis, *Hypersonic: The Story of the North American X-15* (North Branch, MN: Specialty Press, 2002).

Adams plainly recognized that he was susceptible to vertigo. Therefore, at most this "vertigo" (until high accelerative forces were involved) was a contributing factor—rather than an incapacitating cause—drawing away further cognitive resources and capacity in an already task-saturated environment.

Flight Instrument Display and Loss of Mode Awareness

A detailed analysis of reconstructed events using telemetered data along with film from an onboard cockpit camera revealed that the instruments in the cockpit were reading normally and that the airplane was manually put into a rightward flat spin as a result of pilot action. These facts drew attention to the importance of the primary flight display.

The primary flight instrument for orientation—the attitude indicator—was a ball with two bars, or needles, one vertical and one horizontal. Normally, the horizontal bar indicated the aircraft's AOA (also known as alpha) and the vertical bar indicated sideslip (yaw, or beta). On Adams's X-15, the attitude indicator was configured such that it could be put into an alternate mode that would then provide precise roll indication.[36]

After engine shutdown, Adams appropriately turned on the Precision Attitude Indicator switch so that the vertical bar indicated precision roll instead of sideslip. Adams planned to switch the indicator back to its original mode following a wing-rocking maneuver, and well before the time of atmospheric reentry.

The primary flight control instruments are located in the upper center console of the X-15 instrument panel. The row of lights at the top provides the status of the MH-96 adaptive flight control system. The horizontal tape below the attitude direction indicator ball indicates sideslip. (NASA)

Postaccident analysis revealed that when Adams began to use the left-hand ballistic controller, he actually made several inappropriate yaw inputs that caused the aircraft's heading to deviate from the planned flightpath. When investigators correlated the timing of these ballistic control system inputs with indications on the primary flight display, they made an interesting discovery. As the vertical bar moved to the right, Adams put in right ballistic input, as this was the normal method of yaw control, i.e., to "bring the nose to the needle." It became apparent that Adams made

---

36. Bellman et al., "Investigation of the Crash of the X-15-3."

the yaw control inputs in response to the position of the vertical needle on the flight display. Although the checklist had called for Adams to turn the Precision Attitude Indicator switch back to its original setting (alpha-beta), thus causing the vertical needle to once again indicate yaw, he apparently failed to do so. Not realizing this, Adams instead read it as if it were a beta (yaw) indicator. As accident investigators put it:

> Normally, the vertical needle or bar on the attitude-ball presentation indicates a sideslip error, as it did during the boost phase of this flight.... However, after boost and from the time Major Adams correctly selected "precision attitude indicators," the vertical bar was presenting roll angle.... The flight plan called for a return to the sideslip (alpha-beta) indication for reentry.... As a result of Major Adams' long training in flying vertical needles as yaw indicators...it is possible that he may have forgotten that the vertical presentation he was flying was indicating roll and flew it as yaw.[37]

This suggests that a combination of pilot workload and negative transfer of training/expectancy may have biased Adams to forget, or not be fully aware, that the vertical needle was now indicating roll and not yaw. From a Human Factors Analysis and Classification System (HFACS) standpoint, this general mistake was due to technological (display ambiguity) issues, the condition of the operator (adverse mental states and/or physical/mental limitations), and unsafe acts caused by errors that were perceptual, skill-based, and decision-related in nature.

How such a mistake could occur is easily understandable. For this particular flight, Adams had trained for about 23 hours in the X-15 simulator. On his prior flights in the X-15, the vertical needle consistently indicated sideslip (yaw), and he always responded appropriately with yaw control inputs to vertical bar indications. Because the vertical needle coincidentally seemed to respond to his yaw control inputs during that point in the mishap flight, it is very likely that Adams did not realize the error until the situation was beyond the point of recovery, if he realized it at all. This negative transfer of past experience/training contributed to his predicament.

There were other indicators on the instrument panel, however, as well as outside visual cues that could have alerted the pilot to the actual orientation of the aircraft. Adams himself had once told an X-15 crew chief that he would believe his instruments before he would believe anything else.[38]

---

37. Ibid., p. 36.
38. Ibid.

Frame 827
10:34:54

Frame 828
10:34:54.4

Frame 829
10:34:54.8

The last three frames of cockpit camera film, taken as the X-15 began to break up, show the instrument panel and the pilot's relative motion as the aircraft approached both the side-load and normal-load limits. The attitude direction indicator (on the upper center panel) shows that the X-15 was in an inverted dive. (NASA)

So, even if he had misinterpreted his primary flight display, why did he not cross-check this indicator with any of the other panel displays? This is where the issue of vertigo takes on significance. As previously noted, it is likely that Adams experienced the oculoagravic illusion during this phase of the flight and was subject to the added workload of sorting out and troubleshooting the malfunctions in the vehicle and its computer. If this was the case, it would explain why he did not cross-check the three attitude indicators (roll, pitch, and yaw) but flew using only the alpha-beta needles.

Matejka noted:

> He presents evidence of…an unusual susceptibility to vertigo, [which] could have played a significant part in the accident. As mentioned in the previous section, during the flight the pilot seemed unaware of a gross heading deviation in spite of three separate correctly reading instruments and external visual cues that should have made him cognizant of this deviation. He apparently was concentrating on a single instrument, the vertical needle of the ADI, as a pilot might do if he were trying to overcome vertigo. If so, and if the pilot forgot he had switched the function of the display on which he was concentrating, his actions could be explained.[39]

Today, this is relatable to channelized attention, or perhaps distraction, with a resulting "loss of mode awareness," to indicate its standing as a specific type of loss of situational awareness.[40] That is, the pilot probably lost awareness as to which mode his instrument was in, and as a result applied inappropriate flight control input. This was a checklist failure item (adverse mental state/decisional in nature), but it could have been prevented with better display design in terms of clarifying what mode the vertical bars were actually displaying at the time.[41]

Display Design Lessons

The Accident Investigation Board concluded that, in the future, a single display should not be used for two different purposes. This recommendation is particularly relevant for flight operations today because digital displays can be

---

39. Ibid., p. 37.

40. M.R. Endsley, "Toward a Theory of Situation Awareness in Dynamic Systems," *Human Factors* 37 (1995): 32–64.

41. Victor Riley, "Reducing Mode Errors Through Design," *Avionics Magazine, http://www.aviationtoday. com/av/categories/commercial/789.html*, March 1, 2005, accessed October 16, 2008.

used for multiple functions and the indication of which mode is in use is a critical factor in display design.[42] Since the time of the X-15 mishap, the discipline of human factors engineering has evolved and matured to the point of presenting a sophisticated analysis of display design. In the intervening 40-plus years since the time of this mishap, 13 principles of display design have been developed and recognized as standards. Though not mathematical equations or absolute design requirements, they nevertheless represent the principles by which displays can be designed to minimize user error.[43]

In short, these principles are grouped into four main categories. The first group is that of Perceptual Principles (legibility, avoidance of absolute judgment, utilizing top-down processing, principle of redundancy gain, discriminability). The second is composed of Mental Model Principles (principle of pictorial realism, principle of the moving part). The third comprises the Attention-Based Principles (minimizing information access cost, proximity compatibility principle, principle of multiple resources). The fourth addresses the Memory-Based Principles (replacing memory with knowledge-in-the-world, principle of predictive aiding, principle of consistency).[44]

Of particular interest in the X-15 mishap are the principles of display discriminability and replacing memory with knowledge-in-the-world and formal checklist items (which are communications and supervisory factors from an HFACS perspective). Discriminability is a component of distinctiveness (that is, the degree to which a symbol can be identified when standing alone). Symbol identification involves such cognitive processes as feature-learning, feature-extraction, attention, and memory effects.[45]

This principle ties in closely with that of replacing memory-based knowledge (known as knowledge-in-the-head) with mode awareness, or knowledge-in-the-world. Simply defined, knowledge-in-the-head requires learning, sometimes a considerable amount. Such knowledge can be used very efficiently, though not easily on the first attempt. By contrast, knowledge-in-the-world substitutes interpretation for learning. The level of ease with which information is interpreted depends on the exploitation of natural mappings and constraints.

---

42. Ibid.

43. Wickens et al., *An Introduction to Human Factors Engineering*.

44. Ibid.

45. Albert J. Ahumada, Maite Trujillo San-Martin, and Jennifer Gille, "Symbol Discriminability Models for Improved Flight Displays," in *SPIE Proceedings*, vol. 6057, paper 30 (Bellingham, WA: SPIE—The International Society for Optical Engineering, January 2006).

Efficient use of this type of knowledge is impeded by the need to find and interpret external information, but it is generally easy to use on the first attempt.[46]

In the cockpit, pilot understanding of the type of information a display is providing at any given time, or even whether it is active, is particularly important. A simple example from modern general aviation illustrates the point. An instrument such as an artificial horizon (attitude indicator) will likely display an orange flag to indicate that it is inoperative. In fact, the instrument itself will likely be "caged," meaning the moving elements will be mechanically immobilized when the device is inoperative. Otherwise, if the instrument were inactive and the elements were permitted to move freely, the pilot could mistakenly assume that the instrument was active and working, relying on the erroneous information presented by the display as if it were valid and reliable. The caging and the orange flag ideally prevent the errant assumption that the artificial horizon is indicating straight-and-level flight, when in fact it is inoperative; the potential for mishap in this instance is obvious. Discriminability and knowledge-in-the-world that does not depend on pilot memory are two elements that enhance safety in this type of situation.

In the case of the X-15 primary flight display, there was no discriminability or other knowledge in the world by which to differentiate the normal alpha-beta mode from the much less frequently used precision-roll-indicator mode. That is, there was no flag, light, or any other indication on the display clarifying that the displayed mode had been changed. When the pilot changed the instrument mode setting, there was no obvious indication of the change. Because the display was in another mode for the vast majority of the time (during periods of peak attention demand and task saturation), the pilot was unaware that the instrument was in an other-than-normal mode. He flew the aircraft using the display as he had done for the vast majority of X-15 flights and simulator training.[47]

The lack of discriminability, negative transfer of training, and loss of mode awareness were certainly not intentional but nevertheless remain a lesson to be learned for the designers of future systems that incorporate multifunction displays. A design by which a flag or source of illumination indicated the alternate mode of operation would have likely indicated to Adams that he had not yet switched back to the alpha-beta mode.

---

46. Barry A. Romich, "Knowledge in the World Vs. Knowledge in the Head: The Psychology of AAC Systems," *Communication Outlook* 16, no. 2 (1994): 19–21.

47. Bellman et al., "Investigation of the Crash of the X-15-3."

Screening Versus Design

Medical screening in the selection of pilots is a practice that had become well established and widely practiced even in the earliest days of aviation.[48] The experiences of aviators in World War I generated a great deal of interest in screening vestibular system function in pilot candidates, particularly with regard to the propensity for an individual to develop spatial disorientation as a result of performing various flight maneuvers.[49] Even so, the underlying thinking regarding the relationship between spatial disorientation and instrument flying was unclear and frequently inaccurate until studies in the 1920s by David Myers, Bill Ocker, and Carl Crane at Brooks Field, TX, demonstrated a correlation between instrument flying and neurovestibular function. Rightly or wrongly, postural and other equilibrium tests were among the primary means of ascertaining medical qualification for flying duties. In the period surrounding World War II, as the logistics of training aviators become more complex, the medical screening of aviators took on added significance in order to select only those candidates most likely to succeed in training. The complexity of such medical selection processes further increased during the selection of early astronauts for the space program.[50] The idea was to use ever more sophisticated measurements of physiology to identify candidates before training who might have a medical deficit rendering them unfit for flight (or astronaut) duties. By the time the space program was in full swing, the line between medical screening for duty selection and that of clinical research data collection was often blurred. It was not always clear whether data gathered in screening tests indicated abnormality or simply reflected the wide range of normal human physiology.

Indeed, a screening test must render a prediction in some reliable way regarding the condition being assessed. That is, there must be a gold standard by which to assess the screening test's efficacy. Only by determining the sensitivity and specificity (that is, false-positive and false-negative rates) of the screening test, as well as the prevalence of the endpoint in the population under study, can one determine the positive predictive value of the test—that is, the likelihood that a positive test finding actually predicts the condition in question.

The Army Air Corps' attempt to develop standards for the postrotatory nystagmus test is a case in point. As noted earlier, there was a requirement that postrotatory nystagmus should have a duration of between 16 and 36 seconds.

---

48. Roy L. DeHart and Jeffrey R. Davis, eds., *Fundamentals of Aerospace Medicine*, 3rd ed. (Philadelphia: Lippincott Williams & Wilkins, 2002).
49. Previc and Ercoline, eds., "Spatial Disorientation in Aviation."
50. Mackowski, *Testing the Limits: Aviation Medicine and the Origins of Manned Space Flight.*

As a result, a large number of pilot candidates were rejected because they did not meet this requirement. Many later joined the air services of other countries that had less rigorous standards, often becoming excellent aviators.[51] Thus, screening for labyrinthine sensitivity by means of measuring postrotatory nystagmus failed to predict which pilots would later develop spatial disorientation affecting flight performance. Once this became apparent, the requirement for specific duration on the cupulogram was dropped. Such was the case by the time this test was applied in the selection of astronaut candidates. As a result, in the case of Adams, the Aerospace Medicine Division at Brooks AFB found him medically qualified for selection as an astronaut in the MOL program.

While it may be tempting to retrospectively apply the cupulogram duration standard in this case and to say that Adams should have been screened from the program before he began, such reasoning does not address the fact that, based on past performance and as noted earlier, this physiological abnormality arguably could be what set Adams apart from his peers as a superior pilot.

Conclusions

A study of the sole X-15 fatality is a lesson in aerospace physiology and human factors display issues such as mode awareness, situational awareness, the need for checklist clarity, flight-test workload considerations, and physiological issues. Certainly vertigo, or more precisely an aspect of the oculoagravic illusion, was involved; but the historical record clearly demonstrates that this was an example of Spatial Disorientation Type II (Recognized), and not Type III (Incapacitating), until near the end. Moreover, it was merely a contributing factor to the mishap, not its cause. The actual cause related, among other things, more directly to Adams's misinterpretation of the primary flight display after having changed its mode of operation to facilitate data gathering in an airborne science experiment. This loss of mode awareness due to lack of mode/display discriminability resulted in Adams applying inappropriate control input to the aircraft, causing it to enter a spin as it reentered Earth's atmosphere.

Interestingly, at the altitude and airspeed that Adams was flying, he had no sensation of forward motion. He was simply too high above the ground to sense his velocity and direction. This made it doubly important that he knew his orientation, relative to his velocity vector, before reentering Earth's atmosphere.

Other factors contributing to the mishap included an electrical malfunction that caused disturbances in the MH-96 adaptive FCS and in other electrical

---

51. P.M. van Wulfften Palthe, "Function of the Deeper Sensibility and of the Vestibular Organs in Flying," *Acta Otolaryngologica* 4 (1922): 415–448, as quoted by Previc and Ercoline in "Spatial Disorientation in Aviation."

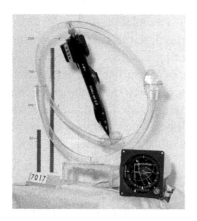

Data from mission telemetry and the cockpit camera film were used to re-create the accident sequence with this model. More than 7,000 frames of film were shot individually, with the model being repositioned each time, to make a stop-motion animation motion picture for use in the accident investigation. (NASA)

components within the aircraft. These malfunctions diverted the pilot's attention, creating an increased workload with probable task saturation, as well as channelized attention. This was apparent to investigators since it was evident that Adams neglected to use instruments effectively during certain phases of flight (which also directly relates to impaired situational awareness). All of these factors, along with the oculoagravic illusion and the lack of mode discriminability, increased pilot workload. This led to the misinterpretation of data reflected in the primary flight display, which led in turn to incorrect pilot input, resulting in a hypersonic spin. Although Adams recovered from the spin, a mechanical limit-cycle oscillation related to the computer system developed, eventually creating aerodynamic forces that exceeded the vehicle's structural limits.

This mishap highlighted the importance of pilot-vehicle interaction throughout the flight. In other words, the way Adams obtained information about the vehicle in order to guide input was as important an issue as his medical fitness for flight or the vehicle's aerodynamic properties. This distinction is not trivial. The laws of physics and the properties of the atmosphere cannot be fundamentally changed; therefore, appropriate individuals who can withstand predicted stresses must be selected as pilots and astronauts. With regard to the human-machine interface, however, opportunities for improvement are not confined solely to selection and training of appropriate individuals. The design of the machine interface can be changed, thereby fundamentally altering the human-machine interface in a way not possible with the human-environment interface. Cockpit design now becomes as important as pilot selection and training with regard to overall mission success.

This investigation of the X-15 mishap highlights the distinction between medical screening and systems design. Further medical screening, as recommended by the X-15 Accident Investigation Board, would have had little significant effect on reducing the risk of recurrence. On the other hand, had the X-15's display design been better and more effective with regard to mode awareness and checklist modification, such improvements would have been significant in addressing loss of mode awareness, the accident's true cause.

Generally, adherence to the principles of display design—particularly with regard to mode awareness—is relevant today in the development of commercial space vehicles undertaking parabolic space flight, as well as in other areas of aerospace operations. In the modern era of transatmospheric operations, intuitive displays that clearly indicate a vehicle's orientation in space, its energy/velocity state, and heading, along with understandable, easy-to-use menus and mode functions, will contribute a great deal to the prevention of future mishaps.

Bruce Peterson flew several of the lifting body research aircraft. He made 3 flights in the M2-F2, made 17 flights in the lightweight M2-F1, and piloted the maiden flight of the HL-10. (NASA)

CHAPTER 5

Six Million Dollar Man

The M2-F2 Task Saturation Mishap

Under the best of circumstances, any pilot faces a significant workload in the cockpit. Ordinary tasks include flying the airplane, navigating, communicating with outside agencies (such as air traffic control), and maintaining situational awareness. When, as in an emergency, demands exceed the pilot's capacity for attention to multiple activities, workload may increase to the point that the pilot experiences task saturation due to an overwhelming number of concurrent demands. In human engineering terms, the pilot's "spare capacity" to sustain additional workload and continue to adequately perform key primary tasks for the situation is unacceptably diminished. As task saturation increases, performance decreases; but due to the cognitive overload, the pilot may not even be aware of the onset of degraded performance.[1]

Military aviators are trained to respond to evident task-saturation by prioritizing. The first priority is effective flight control of the vehicle. As one might expect, test-flying a one-of-a-kind research aircraft at the edge of flight research knowledge frequently entails numerous challenges that can lead to task saturation. One such case involved the M2-F2 lifting body.

Close Calls

By mid-October 1966, three pilots had flown the M2-F2 a total of 12 times. Milton O. Thompson piloted five of those flights before retiring from research flying to transfer into an engineering position as chief of the Research Projects Office at the Flight Research Center. He did not anticipate any future projects as exciting as his previous X-15 and lifting body experience, and he had become bored with the routine proficiency flying required between research flights. "When a pilot gets bored with flying, it is time to quit," he wrote in his autobiography.[2]

---

1. James D. Murphy, *Business Is Combat* (New York: Harper Collins, 2000).
2. Milton O. Thompson, *At the Edge of Space: The X-15 Flight Program* (Washington, DC: Smithsonian Institution Press, 1992), p. 276.

Because the wingless M2-F2 had a low lift-to-drag ratio, the pilot waited until the last possible moment to deploy the landing gear during approach to the lakebed runway. (NASA)

Before leaving the lifting body project, Thompson checked out NASA research pilot Bruce Peterson and Air Force test pilot Donald M. Sorlie in the simulator. On Sorlie's first flight in the M2-F2, he experienced pilot-induced oscillation (PIO). Fortunately, he had learned from Thompson's earlier experience and had planned what he would do if he encountered PIO. The incident also occurred early enough in the flight that he had ample time for recovery.

After two additional flights without incident, Thompson gave approval for Air Force test pilot Jerauld R. Gentry to join the program. Gentry had previously flown the lightweight M2-F1, on two occasions entering a slow roll while still attached to the tow cable. On his first flight in the heavyweight M2-F2, everything went smoothly until just prior to landing. Seconds before touchdown, he discovered to his horror that he was unable to reach the landing gear handle.[3]

---

3. R. Dale Reed and Darlene Lister, *Wingless Flight: The Lifting Body Story* (Washington, DC: NASA SP-4220, 1997).

Northrop engineers had designed the cockpit to accommodate pilots of average height, such as Thompson and Peterson. Consideration had not been given to the needs of shorter pilots with shorter arm spans, and there were no such human-systems interface anthropometric standards in place at that time. Gentry's inability to reach and pull the handle should have been discovered during preflight checkout procedures, but it was not.

As the M2-F2 plummeted toward the ground, Gentry's quick thinking averted tragedy. He loosened his shoulder harness, leaned forward, and pulled the handle. As the landing gear lowered into place, he tightened his harness and executed a safe touchdown.[4]

"The Six Million Dollar Man"

The pilot most often associated with the M2-F2 gained his notoriety in the wake of a spectacular mishap. Grainy film footage of the event ultimately became an iconic part of 1970s pop culture.

A native of Washburn, ND, Bruce A. Peterson grew up in Banning, CA, and attended the University of California at Los Angeles (UCLA) from 1950 to 1953. While at UCLA he held a job as an aircraft assembler for Douglas Aircraft Company.

Peterson enlisted as a naval aviation cadet at Santa Ana, CA, in 1953 and was commissioned a second lieutenant in the U.S. Marine Corps in November 1954. Released from active duty in 1958, he enrolled in California State Polytechnic College at San Luis Obispo, where he received a bachelor of science degree in aeronautical engineering.

Peterson joined NASA in August 1960 as an engineer at the Flight Research Center at Edwards AFB. He joined the Flight Operations branch in March 1962 and was initially assigned as one of the project pilots on the Rogallo Paraglider Research Vehicle (Paresev) program, which evaluated the use of inflatable and noninflatable flexible wings for the recovery of crewed space vehicles. The aircraft resembled a tricycle dangling beneath a hang glider and was towed aloft behind a car or small airplane and released for unpowered landing. Peterson was one of a handful of NASA pilots who made more than 100 Paresev research flights between 1962 and 1964.

Peterson made his first NASA research flight on March 14, 1962, sustaining slight injuries when the Paresev crashed from an altitude of about 10 feet

---

4. Ibid.

during a ground-tow flight. Always the consummate engineer, his first question after impact was, "What happened to the lateral stick forces?"[5]

In preparation for further flight research duties, he attended the Air Force Test Pilot School at Edwards AFB, becoming the first NASA pilot to graduate from the school. Throughout his career, he remained active with the Marine Corps Air Reserve, flying the F9F, OV-10, A-4, and various helicopters.

As a NASA research pilot, Peterson flew a wide variety of airplanes, including the F5D-1, F-100, F-104, F-111A, B-52, NT-33A Variable Stability Trainer, T-33, T-37B, T-38A, C-47, CV-990, Learjet, JetStar, wingless lifting bodies, and numerous general aviation aircraft, as well as several types of helicopters and sailplanes.

As project pilot on the F-111A variable-geometry (swing-wing) jet aircraft, Peterson performed tests related to stability and control, performance, and structural loads. Research with the aircraft included engine inlet and exhaust studies, internal flow investigations, and aerodynamics research.

On December 3, 1963, Peterson flew two flights in the M2-F1 lightweight lifting body. The wooden craft was towed aloft behind a C-47 and released for a gliding landing. Peterson's first flight was uneventful, but the second ended badly when his steep approach resulted in a hard touchdown; the cold weather had thickened oil in the landing gear struts, making them rigid. At touchdown, the M2-F1 came to a sudden stop, and through a cloud of dust Peterson watched his two main wheels bounce across the lakebed and disappear into the distance.[6]

Peterson's first flight in the heavyweight M2-F2, on September 16, 1966, was an unpowered glide flight from an altitude of 45,000 feet. After release from the B-52 mother ship, he executed a 360-degree turning approach and landed on the dry lakebed at Edwards AFB. He made another glide flight in the same vehicle 6 days later.

Peterson piloted the maiden flight of another heavyweight lifting body on December 22, 1966. This was the HL-10 aircraft, which had a somewhat different aerodynamic profile than the M2-F2's. During the 3-minute descent to landing, he discovered he had minimal lateral control over the vehicle. Airflow separation across the control surfaces rendered the HL-10 virtually unflyable, but he managed to land the vehicle safely—a tribute to his considerable piloting skills. As a result of data collected during the nearly disastrous flight, the HL-10 was modified to rectify the problem. It eventually went on to become one of

---

5. Peter W. Merlin, "The Real Six Million Dollar Man—Bruce Peterson," *World X-Planes* no. 3 (summer 2006).

6. Ibid.

the most successful lifting body designs ever produced. In fact, the aircraft's design was a strong contender for that of the final Space Shuttle.[7]

"That Chopper's Going To Get Me"

Peterson's apparently good luck ran out during the 16th flight of the M2-F2 on May 10, 1967, when he strapped himself into the aircraft on its launch pylon beneath the right wing of the B-52. The flight's purpose was pilot proficiency and evaluation of damper systems during maneuvers in order to obtain stability and control data, particularly with regard to the vehicle's lateral-directional characteristics. The flight plan called for launch east of Rogers Dry Lake on a northerly heading; maneuvering to collect data; a turn to the west with a long, straight base-leg for additional maneuvers; and a final turn to the south for a landing on the west edge of lakebed Runway 18. Since it was to be a glide flight, the rocket engine was not fueled. The M2-F2 was, however, equipped with two 500-pound-thrust hydrogen-peroxide rockets to provide emergency lift during landing.

A flightcrew briefing took place on May 9 and included Peterson, as well as B-52 pilot Col. Joseph F. Cotton, copilot Maj. Jerry D. Bowline, launch-panel operator Victor W. Horton, and SSgt. Joseph F. Dillon. John A. Manke was designated to fly chase in an F5D-1, following the lifting body from drop through landing. William H. Dana served as alternate chase in an F-104N, and Lt. Col. Sorlie was assigned to fly photo chase in a T-38. Capt. Gentry, designated ground controller for the flight, presided over the meeting. Among other things, the briefing included a description of planned deployment of a rescue helicopter to hover over the lakebed, east of the runway. Gentry said he would brief the pilot of the H-21 rescue helicopter, who was not present at the meeting, later in the day. The next morning, however, there was a helicopter crew change. The original helicopter pilot then briefed his replacement, Capt. Larry D. McLaughlin, who flew the rescue helicopter for the mission.[8] Gentry, who was assigned as test conductor, later briefed the rescue helicopter pilot.

As the B-52 took off carrying the M2-F2 and climbing to an altitude of around 44,000 feet, McLaughlin flew the H-21 along the south side of the runway for the benefit of a photographer on board. He then proceeded to Runway 18 to inspect the condition of the airstrip. On previous lifting body missions, the helicopter had hovered about 1,000 feet over the intersection of runways 18 and 23. For

---

7. Ibid.

8. Donald R. Bellman et al., "Investigation of Landing Accident with the M2-F2 Lifting Body Vehicle on May 10, 1967, at Edwards, California," NASA Flight Research Center, Edwards, CA, June 1967.

this flight, however, planners decided to allow McLaughlin to land on the lakebed and remain there until the M2-F2 was launched.

System checks went smoothly with the exception of poor reception in both of the lifting body's communications radios. The problem was not serious enough to cause the flight to be terminated, however. During the climb, Peterson announced that he was planning to change his landing path—angling across the lakebed runway—in order to reduce crosswind effects.

At launch time, the lifting body dropped away from the B-52, and Peterson maintained a level attitude and 15 degrees AOA for the first pitch maneuver. He was pleasantly surprised to find that the vehicle behaved much as it had in the simulator. After the first turn, he executed a gentle pushover and trimmed the aircraft in preparation for test maneuvers.

Peterson noted that the aircraft was flying smoothly, although he found the turn rate somewhat better than it had been in simulation. He turned down the dampers and performed a pitch doublet. Deciding that the effective aileron-rudder interconnect was higher than had been simulated, he lowered his stability augmentation system gains.

While maneuvering, Peterson noted some controllability problems and lowered his stability augmentation system gains. Generally, all went well until the final turn toward the lakebed, when the wingless craft began a violent "Dutch roll" motion. "Boy, there's some glitches," he exclaimed.[9]

Dutch roll is an aerodynamic phenomenon in which a roll (longitudinal input) results in not only roll, but also a change in the nose position, or lateral deviation of the nose of the aircraft. This is an example of multi-axis coupling, which should be fairly minimal for optimal controllability.

Once the lifting body had commenced its glide, McLaughlin took off in the helicopter. He positioned the H-21 to the east of the runway and hovered at an altitude of about 10 feet. This would soon present Peterson with an additional distraction.

Strong winds blew the M2-F2 to the north, necessitating corrective maneuvers, but Peterson had already planned for this, announcing his intention to land at an angle across the runway in order to mitigate crosswind effects. As he descended through 16,000 feet, Peterson experienced radio difficulties. He could hear Manke's garbled transmissions but could not discern what the chase pilot was saying.

With less than 7,000 feet of altitude remaining, Peterson entered the final turn while slightly increasing stability augmentation system gains. Without warning, he experienced a very high divergent rolling motion at more than

---

9. Ibid.

Observers in chase planes, including the F-104 (left) and the F5D-1, kept a close eye on the M2-F2 during each flight. (NASA)

220 degrees per second. Fighting against disorientation, he attempted to damp the motion by increasing the AOA.[10]

The horrible rolling motions prompted the chase pilot to advise Peterson to eject: "OK, buddy, when she rights up, get out!"

Peterson regained control but found his heading had altered considerably. He looked down, disoriented and surprised because he could not see any runway markings on the lakebed. He thought he was headed southeast but had insufficient altitude and energy to make any corrective actions to get back toward the runway, and he did not have any kind of markings that would aid his depth perception. The featureless playa provided no clues.

When he looked up, Peterson saw that the rescue helicopter appeared to lie directly in his flightpath. He radioed, "Get that chopper out of the way."[11]

Manke advised him to prepare to lower the lifting body's landing gear, but Peterson was still worried about the H-21. "He's all right," called Manke, "You're okay, you're clear, watch your gear, Bruce."

---

10. Ibid.

11. Ibid.

As Peterson fought to maintain control and execute his landing flare, he continued to worry that the rescue helicopter appeared to be in his path. After Manke advised Peterson to lower his landing gear, Peterson acknowledged the call but added: "That chopper's going to get me, I'm afraid."

As soon as McLaughlin heard the first call, he climbed to 30 feet, turned to the south, and accelerated to 60 knots to get away from the runway. The M2-F2 missed the helicopter by several hundred feet. Peterson now fired the landing rockets, flared, and lowered the landing gear. Although less than 2 seconds was required for locking and lowering the landing gear, it was too late.[12] Before the gear was fully down, the M2-F2 struck the ground at an estimated 250 mph. It bounced, tumbled, and rolled across the lakebed in a cloud of dust, finally coming to rest upside down.

McLaughlin turned the helicopter back toward the stricken research aircraft. Jay L. King, of the Operations Division, and Capt. Michael Hall, an Air Force flight medical officer, jumped off the helicopter and ran to Peterson's aid. Peterson was found still strapped into the cockpit, with his head and arms hanging down. He began to moan, and responders realized he was still alive, though badly injured with extensive facial trauma. The responders unbuckled his seat belt and slid him downward, though he was still restrained by his parachute and seat pack. Additional rescuers, who by now included the fire chief and pilot aide Joe Huxman, worked to extricate Peterson from the remaining cockpit encumbrances. They placed him on a stretcher and brought him to the waiting helicopter. He was rushed to the Edwards AFB hospital for initial medical stabilization, then transferred to the hospital at March AFB in Riverside, CA, for further treatment before eventually being sent to the UCLA Medical Center in Los Angeles, CA, for definitive care.[13]

From the Ashes

Following an extensive accident investigation, the M2-F2 was not scrapped, as might have been expected considering the damage it had sustained. Instead, it was returned to Northrop Aircraft Corporation in Hawthorne, CA, for a comprehensive inspection that lasted 60 days. There, Northrop technicians placed the battered vehicle in a jig to check alignment of the airframe and removed the external skin and portions of the secondary structure.[14]

---

12. Merlin, "The Real Six Million Dollar Man."

13. Ibid.

14. Richard P. Hallion and Michael H. Gorn, *On the Frontier: Experimental Flight at NASA Dryden* (Washington, DC: Smithsonian Books, 2003), p. 157.

At the same time, NASA engineers conducted a series of wind tunnel tests to determine whether structural modifications could improve the vehicle's stability. Though conventional winged aircraft are subject to roll, spiral, and Dutch roll, the wingless lifting bodies can also experience a unique motion known as a coupled roll-spiral mode. During 16 glide flights in the M2-F2, severe lateral PIO occurred on three occasions during final approach to the runway. The Flight Research Center's Robert W. Kempel made a qualitative analysis of data collected during these events using recorded time histories of each flight and pilot comments on each maneuver. He then performed a systems analysis using predicted aerodynamic stability and control derivatives for the flight conditions at which Peterson's PIO occurred in order to determine the root cause of the oscillations. Kempel's results directly related preflare, low-angle-of-attack PIO tendencies to the formation of a coupled roll-spiral mode that caused Peterson to generate a closed-loop lateral instability, which was further aggravated by attempts to coordinate aileron roll with the rudders.[15]

The most satisfactory solution appeared to be the addition of an extra vertical stabilizer, centrally mounted between the two outer fins at the aft end of the fuselage. In March 1968, NASA's Office of Advanced Research and Technology authorized Northrop to structurally restore the M2-F2 and return it to the Flight Research Center. While NASA officials debated the vehicle's future, engineers determined the characteristics of a configuration with three vertical fins that came to be known as the M2-F3. The proposal to modify the craft was approved in late January 1969, and by June 1970 it was flying again. Benefiting from lessons learned in Peterson's mishap, the M2-F3 was flown 27 times, reaching a maximum speed of 1,064 mph and a peak altitude of 71,500 feet. The M2-F3 was retired in December 1972 and eventually placed on display in the Smithsonian Institution's National Air and Space Museum in Washington, DC.[16]

Peterson eventually recovered sufficiently—despite losing sight in one eye due to a secondary infection while in the hospital—to fly NASA support missions and occasional research flights and to resume his Marine Reserve flying duties. He continued to fly for NASA until 1971, performing research missions in the T-33, F-104B, F-111A, CV-990, and Aero Commander. He also flew NASA's SH-3A helicopter. The Marine Corps gave him a waiver that allowed him to fly with a copilot, and he continued to fly the OV-10 airplane, as well as

---

15. Robert W. Kempel, "Analysis of a Coupled-Roll-Spiral Mode, Pilot Induced Oscillation Experienced with the M2-F2 Lifting Body" (Washington, DC: NASA TN D-6496, 1971).

16. Hallion and Gorn, *On the Frontier*, pp. 157–158.

The wreckage of the M2-F2 lies on its back on Rogers Dry Lake. Bruce Peterson's helmet is visible in the foreground. (NASA)

the AH-1G and CH-46 helicopters. During his flying career, Peterson logged more than 6,000 flight hours in nearly 70 different types of aircraft.[17]

Peterson gained a small measure of fame when his accident and subsequent recovery inspired a 1970s television series called *The Six Million Dollar Man*. The storyline featured a test pilot who, having been injured in the crash of a lifting body vehicle, is rebuilt with advanced "bionic" technology. Film footage of the M2-F2 accident was used in the show's opening credits.

After giving up flying, Peterson served as research project engineer on the F-8 Digital Fly-By-Wire program of the late 1960s and early 1970s. He later assumed responsibility for safety and quality assurance at Dryden until his retirement in 1981.

He left NASA for a position with Northrop, where he assumed responsibility for safety and quality assurance in the testing of the B-2 Advanced

---

17. Merlin, "The Real Six Million Dollar Man."

Technology Bomber. From 1982 until 1994, Peterson worked in Northrop's B-2 division at Air Force Plant 42 in Palmdale, CA, and at Edwards AFB, becoming manager of system safety and human factors.[18]

Analysis of the M2-F2 Mishap

Several significant human factors contributed to the M2-F2 accident. These included habit pattern transfer, spatial disorientation, distraction, and task saturation.

Habit Pattern Transfer: "Instinctive" Rudder Input

The M2-F2 Accident Investigation Board noted:

> The pilot's use of rudder during the low angle-of-attack condition aggravated the lateral oscillations and increased the time that the vehicle was out of control. Rudder was used although the pilot was fully aware that rudder inputs should not be made under these conditions. However, the use of rudder to coordinate lateral control is almost instinctive. On two previous flights, two other M-2 pilots also made rudder inputs under similar circumstances, even though they, too, were aware that such inputs (even very small amounts of rudder) were not desirable.... It is the opinion of the Board that the uncontrolled lateral oscillations contributed to the accident.... After the accident, the pilot was queried about his use of rudder control. He said he was unaware that he had used this control, but frankly admitted that he might have, inasmuch as it was almost instinctive with him.[19]

Muscle memory, otherwise known as motor memory, was described earlier in the discussion on the first flight of the M2-F2, in which Milt Thompson experienced severe lateral oscillation due in part to a difference in flight controls between those in the simulator used for training and those installed in the actual research vehicle. Procedurally, muscle memory is required for tasks utilizing motor skills, such as riding a bicycle. Once learned, these skills are rarely lost, even if not used for some time. Procedural muscle memory involves the proprioceptive senses, and corresponds to "how-to-do-it" knowledge—i.e., the actual motions involved in accomplishing a given task. Examples include executing a golf swing, playing a particular piece of music on the piano, or

---

18. Ibid.

19. Bellman et al., "Investigation of Landing Accident with the M2-F2," pp. 6–8.

performing an emergency procedure such as an aircraft ejection sequence. This is sometimes called habit pattern and can be particularly important when transferring from the operation of one system to that of another.[20] For example, the pilot of an F-4 who transitions to the F-16 would be required to learn the use of a different pattern of procedures for initiating an ejection sequence (e.g., pulling forward on the "ear handles" located above and behind the pilot's head in the F-4 versus pulling upward on a D-ring between the pilot's knees in the F-16). Thus, in Thompson's case, rudder input leading to lateral oscillation was an unanticipated result of habit-pattern transfer from flight procedures of virtually all other aircraft. In this case, it was seen even in a very accomplished test pilot when flying an extremely unconventional, overly sensitive, and barely stable aircraft.

Spatial Disorientation

According to the M2-F2 Accident Investigation Board:

> The pilot was undoubtedly disoriented when he recovered from the uncontrolled lateral oscillation very near the ground. At this point, he was committed to land in a new area on the lakebed close to the rescue helicopter. There is no question that the pilot was greatly distracted and concerned over a possible collision with the helicopter. Whether or not he should have been able to regain his senses and to have overcome the distraction in time to have made a good landing is difficult to decide. Dr. James Roman [Chief, Biomedical Program Office] thought that the violent lateral oscillations could have disoriented the pilot for a period of 10 seconds…. Also, the landing away from the customary visual reference markings would require additional time for the pilot to adapt in order to make reasonable height judgments. Since there were only 14 seconds between the conclusion of the lateral oscillations and the touchdown, it can be concluded that the pilot performed as well as could be expected in this area.[21]

---

20. Richard C. Atkinson and Richard C. Shiffrin, "Human Memory: A Proposed System and Its Control Processes," in *The Psychology of Learning and Motivation: Advances in Research and Theory*, vol. 2, ed. K. Spence and J. Spence (New York: Academic Press, 1968).

21. Bellman et al., "Investigation of Landing Accident with the M2-F2," p. 8.

Furthermore, in reference to the Dutch roll lateral oscillations, Peterson, in his report to the Accident Investigation Board, stated, "It was disorienting."[22] Given these remarks, there is also an element of a loss of situational awareness at the end of the flight that was due to high controllability workloads and being task saturated.

Spatial disorientation refers to the effects of motion on the vestibular apparatus of the inner ear, specifically the effect upon the fluid-filled semicircular canals that act as the brain's sensors for angular acceleration. Within the semi-circular canals, movement of an oily fluid across specially positioned hair cells in the inner ear allows the brain to detect rotational motion. When the head rotates, the fluid at first tends to remain stationary, causing the hair cells to bend in a particular direction; this is sensed by the brain as a particular rotational movement. As the motion continues, the fluid will achieve the same rate of movement, and resulting perception of the motion, in the absence of other visual or proprioceptive cues, will be lost. However, when the motions cease, the fluid will continue to move, creating a perception of motion in the opposite direction by the brain. When this occurs during flight, the result can be disorienting to the pilot, who must rely exclusively during this period of time on visual cues to maintain his understanding of which way is up. This was compounded by a lack of cues enabling depth perception, as Peterson noted during the postaccident investigation interview:

> When I looked down, I was disoriented and was surprised because I didn't see any runway markings at all and I had the distinct feeling that I was heading SE, but that altitude and energy were too low to make any corrective actions from that point on to get back towards the runway or any kind of markings for aid in depth perception.[23]

The perception of depth by the eye and the brain relies on two different processes. There are the binocular cues of depth perception, to include oculo-motor convergence or divergence, as well as retinal disparity, otherwise known as stereopsis. The binocular cues, however, are much less effective and relevant at distances beyond approximately 30 meters. At that distance, the monocular cues of depth perception assume much greater importance. These cues include size and shape constancy, linear perspective, light and shadow, depth of focus, image overlap (occlusion), and motion parallax. It is beyond the scope of this

---

22. Ibid., p. 5-5.
23. Ibid.

narrative to go into the details of each of these cues, but the following description from the mishap report clearly illustrates the inherent hazards faced by Peterson as he approached the lakebed in the M2-F2:

> The bed of Rogers Dry Lake at Edwards on which the research vehicles are landed is a relatively smooth, level, glaring surface that stretches for miles in all directions. When approaching the lake in preparation for a landing on it, pilots almost universally find it difficult to judge their height because of the lack of visual references of known size on which to base a judgment. Pilots have a similar effect when flying close to a smooth-surfaced body of water. Consequently, the commonly used runways are marked with broad, black tar strips even though much greater areas are quite suitable for landing. It has been Flight Research Center policy to always make landings of unpowered research aircraft, such as the lifting bodies and the X-15 vehicles, close to these tar strips to provide depth-perception reference, and to have chase pilots call out the height of the research vehicle above the lakebed as the flare is completed and the pilot "feels" for the ground.
>
> On this particular M-2 flight, the loss of lateral control just before the flare maneuver prevented the pilot from completing the S-turn and caused him to land a significant distance east of the marked runway rather than close to the runway…. Thus, there were no significant ground markings in the area of the final flare and touchdown, and the pilot's depth perception would have been impaired…there was also a lack of the customary callout by the chase pilots.[24]

Hence, after the experience of the lateral oscillations, Peterson was spatially disoriented with regard to his height above the ground. This occurred at a point in the flight when the timing and sequence of events were becoming more critical. But there were to be much greater human factors with which to contend, including those of distraction and additional cognitive workload forced by the helicopter he perceived to be in his flightpath.

Attention and Distraction: The "Helicopter Problem"

At a critical point during the final approach for landing, Peterson became distracted by the presence of the rescue helicopter. The results were described in the accident report:

---

24. Ibid., p. 8.

The M-2 pilot did not actually see the helicopter until after he recovered from the loss of lateral control about 14 seconds before touchdown. At this time, the helicopter was nearly directly in front of the M-2 because of the inadvertent heading change as a result of the lateral oscillation. The helicopter looming up ahead at a time when the M-2 was essentially committed to a landing path came as a shock to the M-2 pilot, particularly because he had not envisioned the helicopter being over the lakebed at all. His great concern over this matter is evidenced by his three radio transmissions and the manner in which these transmissions were made. This added concern at critical time in the landing flare undoubtedly detracted from the pilot's ability to make a good landing and is considered one of the major causes of the accident.[25]

This was confirmed during the postaccident investigation interview with Peterson, who stated the following:

> Also at this point and time I saw a helicopter in front of me and was extremely concerned that holding my present course, and were he to remain at his present location that we were on a collision course.[26]

Similarly, in chase pilot Manke's statement to the Accident Investigation Board, he noted:

> The helicopter was definitely in front of us.... I realized then that the distraction of the helicopter could cause some serious problems because during the final portion of the M2 flight the pilot has a lot to think about and a lot to get done.[27]

This sequence of events clearly demonstrates the principles of attention and cognitive tunneling. Along these lines, cognitive workload is closely related to attention, and attention is closely related to vision. To better understand the dynamics of resources available for attention and workload, it may be helpful, first of all, to understand some elements of human visual-information processing.

---

25. Ibid., p. 12.
26. Ibid., p. 5-5.
27. Ibid., p. 5-9.

The retina of the eye is literally an extension of sensory brain tissue. The retinal cells are similar in many ways to the neurons of the brain, and visual-information processing literally begins at the level of the retinal cells. That aspect of the retina processing peripheral vision is composed primarily of so-called rod-shaped cells. These cells have very low acuity in bright light, or daytime operations, but are more sensitive to lower levels of light than the central vision. The cells of the peripheral vision are well-adapted to detecting signals—e.g., flashing lights or motion—and to maintaining a sense of visual balance—e.g., perception of a flat or sloping horizon. In contrast, the cells of the central (foveal) vision are more cone-shaped in microscopic appearance. Likewise, they are specially adapted to color vision in brighter light, and their higher degree of concentration within the retina makes for higher visual acuity.

In general, the peripheral vision detects motion or movement (or flashing lights) and serves as a signal for the brain to move the eye such that the object of interest is focused on the central, foveal vision. Objects focused within the central vision are by and large the objects of focused attention, and it is objects within the central vision that are generally the objects of conscious processing, otherwise known as attention.

Attention may be considered the concentration of the mind on a single object or thought. Psychologists usually divide attention into three main types: selective, divided, and focused (or sustained). Selective attention refers specifically to the act of directing cognitive processing to a limited range of information while at the same time often ignoring other sources of information. For example, one may attend to the color of an object, but not its shape. Divided attention refers to the allocation of available cognitive resources to multiple tasks executed simultaneously, with the ability to attend to more than one type of information at the same time. For example, a person may carry on a conversation while simultaneously driving a car. Finally, focused (or sustained) attention recognizes the fact that only one thing can truly be the subject or focus of attention at any one time. Focused attention may be voluntary or involuntary, and several factors affect what becomes the focus of attention, including meaningfulness, salience (i.e., structure of display), color or intensity, and modality.

According to the single resource theory of attention, there is a single, undifferentiated pool of attentional resources available to all tasks and mental operations. The theory holds that cognitive focus is analogous to a searchlight beam, and as the searchlight moves, everything within the beam is processed, voluntarily or otherwise. In this model, selective attention is generally viewed as a serial-processing activity, whereas divided attention is more or less a parallel-processing activity. Hence, divided attention—the ability to monitor multiple channels at once—predicts situations in which there will be an inability to adequately divide attention between two or more equally demanding sources of information, such as flying an aircraft on instruments with an added emergency task. The performance resource function is the allocation of attentional resources when multiple tasks are competing for attention and maximum single-task performance is at 100 percent resource allocation. When there is difficulty isolating a single channel of high priority, this allocation can lead to a failure of focused attention, otherwise known as distraction.

Divided attention, or multitasking, thus becomes the rapid rotation of the searchlight beam across two or more tasks. Becoming absorbed in a single task or focused on a single piece of incoming information is known as cognitive tunneling. Also called channelized attention, this is selective attention focused on one area to the exclusion of other information of importance for proper decision making; in effect, the attentional searchlight stops moving back and forth. In this case, things that are more likely to draw away one's attentional resources are those that are difficult, novel, interesting, or emotionally charged.

Another way of thinking about attentional resources is that of the "radar scanner" hypothesis of attention. In this model the spotlight of attention, the attentional radar beam, sweeps around the areas of consciousness at various speeds and intensities to optimize task performance—crudely similar, for illustrative purposes, to when an air traffic radar beam sweeps around the sky to locate aircraft in the vicinity. When a particular area needs more attention, then an attentional "sector scan" is instituted by the mind's executive controller; but the focus of attention on this sector comes at the expense of paying attention to the other sectors during this focused effort.[28]

The best pilots and information-processing managers know when and how to properly pay attention to competing interests, thereby optimizing the allocation of attentional resources and having a chance to attain the best possible situational awareness for the circumstances at hand. At this point, it is helpful to understand some basic principles of memory function.

Cognitive psychologists generally describe memory using a three-stage model: sensory memory (i.e., the sensory store, or perception of incoming information), short-term memory (otherwise known as working memory), and long-term memory.[29] The sensory store is of short duration—less than 1 second for visual and tactile information, about 2 seconds for auditory information. It is of unlimited capacity, with voluntary control of information selected for fur-

28. Dwight A. Holland, "Peripheral Dynamic Visual Acuity Under Randomized Tracking Task Difficulty, Target Velocities, and Direction of Target Presentation" (Ph.D. diss., Virginia Polytechnic Institute and State University, 2001).

29. Atkinson and Shiffrin, "Human Memory: A Proposed System and Its Control Processes."

ther processing. Overall, the sensory store leads to the perception of the sensed information. The short-term, or working, memory provides resources to retrieve and maintain information during cognitive processing. Short-term memory capacity is limited, and most individuals can hold five to nine (typically about seven) items simultaneously in working memory.[30] Finally, long-term memory consists of that information stored permanently in the brain. This storage takes time and effort but is of almost unlimited capacity and duration. For the purposes of the discussion here, long-term memory will not be discussed at length.

Awareness of a situation in the outside world relies upon appropriate sensory organ function, with appropriate processing necessary to perceive the incoming information. However, the selective attention will identify those parts of the incoming information that are deemed most relevant for the situation. Based on input from the long-term memory (including habit pattern formation), as well as the input from the short-term memory (working memory), the brain will utilize the selected incoming information to build a mental model of the situation. As a result, a particular response will be selected and then executed, with appropriate feedback on the state of the situation in the form of newly incoming information, such that the process continues in a recurring fashion.[31]

Attention and Workload: Landing Flare

As can be assessed from the foregoing discussion, Peterson's distraction caused by the helicopter—in the context of spatial disorientation with regard to height caused by the lateral oscillations and the lack of depth cues—ultimately led to the slight delay in lowering the landing gear, which was what eventually led to the mishap. The Accident Investigation Board saw a possible solution:

> Furthermore, it is believed that a modified approach pattern with more time and altitude allotted to the pilot for positioning and landing would help to eliminate the need for entering the marginal control region close to the ground.[32]

---

30. G. Miller, "The Magical Number Seven, Plus or Minus Two: Some Limits on Our Capacity for Processing Information," *Psychological Review* 63, no. 2 (1956): 81–97.

31. Christopher D. Wickens, John D. Lee, Yili Liu, and Sallie E. Gordon Becker, *An Introduction to Human Factors Engineering*, 2nd ed. (Upper Saddle River, NJ: Pearson Prentice Hall, 2004), p. 122.

32. Bellman et al., "Investigation of Landing Accident with the M2-F2," p. 7.

Peterson corroborated this analysis in the postaccident investigation interview:

> I don't actually remember throwing the bypass switch and that was the switch I intended to use, was the bypass switch, but I do remember having my hand on the switch and I wasn't going to move my hand from the handle until I lit the landing rocket, so I assumed that I did, and of course, the next thing I did was go to the gear handle…but I do remember putting the gear down and almost instantaneously after extending the gear, I experienced a very high deceleration and roll acceleration.[33]

Within the framework of the information-processing model, it is clear that the workload was remarkably excessive, combined with a very distracting element. It seems remarkable that the pilot was able to function as well as he did under the circumstances.

The ultimate failure in execution resulted from task saturation of available attentional resources. This increased workload was exacerbated by a lack of depth perception with regard to height, as well as distraction due to the imminent danger of death by collision with another aircraft—in this case, the helicopter.

The human factors leading to this situation were not only those of the pilot in the cockpit; they extend to the change in the planned flightpath—which exacerbated the change in the actual flightpath—combined with the change in position of the helicopter before the flight sequence. Inadequate briefing of the helicopter pilot compounded the problem.

Conclusions
As flight research has progressed to entail increasingly complicated aerospace vehicles flying at higher altitudes and speeds (and at times featuring unusual aerodynamic properties, such as those of the lifting body concept), there have been corresponding increased demands placed on the cognitive resources of the pilot—the human in the loop—in order to maintain stability and control in even the best of flight conditions. When unanticipated events occur, as they often do in the flight-test and research environment, the demands of controlled flight can exceed the resources of even the most highly selected and well-trained humans. By the late 1960s, flight research vehicles had reached an extremely challenging level of complexity. Subsequent development of computerized FCSes has helped mitigate pilot workload with regard to the basic elements of stability and control.

---

33. Ibid., p. 5-5.

Raptor 4008, the eighth F-22A engineering and manufacturing development airframe, streaks through the skies over California's Mojave Desert. The aircraft incorporated revolutionary advances in airframe structures, materials, low-observable technology, propulsion systems, maneuverability, and integrated avionics. (U.S. Air Force)

CHAPTER 6

Almost-Loss of Consciousness in the F-22A Raptor

A pilot in a maneuvering aircraft is subjected to centripetal acceleration opposed by equal and opposite inertial or centrifugal force. In aerospace medicine, inertial force on the restrained pilot is quantified in multiples of the normal acceleration due to gravity (9.82 meters per second squared) and described in dimensionless units of g-forces. Pilots performing high-g maneuvers, such as those involved in aerial combat and some research flying, experience acceleration forces that affect the body's cardiovascular, pulmonary, musculoskeletal, and nervous systems, causing problems that affect aerospace safety.

In order to define acceleration effects, aerospace physiologists have developed nomenclature symbolizing the physiological effects that result from inertial forces on the three axes of the human body. The inertial force, or $+G_z$, is produced from head to foot, as when the aircraft maneuvers in a tight inside turn. Force in the foot-to-head direction, $-G_z$, results from an outside turn. Transverse and lateral g-forces are referred to as $\pm G_x$ and $\pm G_y$, respectively.[1]

Human performance in the cockpit may be adversely affected by exposure to acceleration that induces altered states of awareness ranging from grayout to brief blackout to catastrophic g-induced loss of consciousness (G-LOC). These are caused by the differential between the location of the central nervous system and the heart within the $+G_z$ field when blood is pushed away from the pilot's brain, through the chest, and into the legs and feet. With increasing acceleration in the $+G_z$ direction, the heart must generate higher driving pressure to sustain adequate perfusion to the brain. Neurological tissues become ischemic, resulting in spatial disorientation or loss of situational awareness.[2]

---

1. James E. Whinney, Ph.D., M.D., Aeromedical Research Division, Civil Aerospace Medical Institute, "Sustained Acceleration Exposure," section II.2.7, *Advanced Aerospace Medicine On-line*, http://www.faa.gov/other_visit/aviation_industry/designees_delegations/designee_types/ame/tutorial/, September 29, 2005, accessed February 18, 2010.
2. Ibid.

Whether in flight testing or air combat, high-g maneuvering requires a pilot to maintain high levels of spatial orientation. Not surprisingly, the hazards of loss of situational awareness (also known as spatial disorientation) have increased with advancements in aircraft maneuverability. Improved capabilities for flight at high AOA, unusual maneuvers, and high-g turns impose greater demands on human physiological and cognitive functions.

One of the most insidious conditions leading to altered states of awareness, or inability to orient oneself spatially, is known as almost-loss of consciousness (A-LOC). Similar to G-LOC, this condition is induced by $+G_z$ acceleration stress that often occurs with short-duration, rapid-onset g-exposure. It is characterized by deficits in the pilot's cognitive and motor functions, but without complete loss of consciousness. The resulting neurocognitive symptoms may result in loss of situational awareness for several seconds. Pilots experiencing A-LOC have reported brief and variable episodes of confusion, amnesia, apathy, spatial disorientation, weakness, and twitching of the hands.[3]

Aircraft accidents attributed to loss of situational awareness are of major concern in combat flying, particularly as more advanced fighter aircraft designs are introduced. The air combat maneuvering environment is characterized by frequent and repetitive excursions to high $+G_z$ levels over several minutes, on average, during an engagement. The resulting stresses on the pilot can lead to reduction in $+G_z$ tolerance, increasing the risk of A-LOC or G-LOC.[4]

A tragic example of this occurred during the testing of a Lockheed Martin F-22A Raptor at Edwards AFB in 2009.

Advanced Tactical Fighter

The F-22A evolved from air combat studies carried out in the late 1970s and early 1980s that culminated with the Advanced Tactical Fighter (ATF) program. Taking advantage of advances in FCSes, composite materials, lightweight alloys, high-power propulsion systems, and stealth technology, Air Force officials sought to develop a combat aircraft for the 21st century. The ATF was to replace the F-15 as a high-performance, long-range air-superiority fighter. Two teams of contractors built technology demonstration prototypes for a fly-off competition that began in 1990. In April 1991, the Air Force ended the competition with the selection of Lockheed's YF-22 as the winner. Lockheed

---

3. Wg. Cdr. A. Sinha and Wg. Cdr. P.K. Tyagi, "Almost Loss of Consciousness (A-LOC): A Closer Look at Its Threat in Fighter Flying," *Indian Journal of Aerospace Medicine* 48, no. 2 (2004): 17–21.

4. Sophie Lalande and Fred Buick, "Physiologic $+G_z$ Tolerance Responses over Successive $+G_z$ Exposures in Simulated Air Combat Maneuvers," *Aviation, Space, and Environmental Medicine* 80, no. 12 (December 2009).

merged with Martin Marietta in 1995, a year after the start of full-scale production of the F-22A.

The production configuration incorporated revolutionary advances in airframe structures, materials, low-observable technology, propulsion systems, maneuverability, and integrated avionics. It was designed to penetrate enemy airspace and achieve a first-look, first-kill capability against multiple targets. Nicknamed the Raptor, the F-22A featured aerodynamic and powerplant characteristics allowing supersonic cruise without use of afterburner.[5]

The F-22A features a state-of-the-art "glass cockpit" in which traditional flight instruments have been entirely replaced by electronic liquid-crystal display monitors. In order to reduce pilot workload, multifunction displays serve as flight instrumentation and provide information for communication, navigation, and identification as well as integrated caution, advisory, and warning-system data.

A side-stick controller located on the right console is force-sensitive, with a throw of only about one-quarter of an inch. Throttles are located on the left console. To support pilot functional requirements, buttons and switches on the grips are coded by both shape and texture to allow control-by-feel of more than 60 different time-critical functions.

Cockpits of earlier fighters were sized to accommodate 5th- to 95th-percentile pilots (a range of only 90 percent). The F-22A cockpit, however, accommodates 0.5th- to 99.5th-percentile pilots (the average body size of 99 percent of the Air Force pilot population). Additionally, the rudder pedals are adjustable, and the pilot has 15-degree over-the-nose visibility as well as excellent over-the-side and aft visibility.

For cases of serious emergency, the F-22A is equipped with an improved version of the ACES II ejection seat, predicted to provide a survivable egress at speeds of up to 600 knots. An active arm restraint system eliminates arm-flail injuries during high-speed ejections. Other improvements include a fast-acting seat stabilization drogue parachute to provide increased seat stability and a new electronic seat/aircraft separation sequencing system. Additionally, the 360-pound canopy is weighted slightly more on one side to reduce the chance of post-ejection collision between the canopy and the pilot's seat.[6]

---

5. David C. Aronstein, Michael J. Hirschberg, and Albert C. Piccirillo, *Advanced Tactical Fighter to F-22 Raptor: Origins of the 21st Century Air Dominance Fighter* (Reston, VA: American Institute of Aeronautics and Astronautics, 1998).

6. "F-22 Raptor Cockpit," GlobalSecurity.org, *http://www.globalsecurity.org/military/systems/aircraft/f-22-cockpit.htm*, accessed March 20, 2010.

Developmental Testing of the Raptor

Lockheed Martin built nine engineering and manufacturing development (EMD) airframes for developmental testing (serial nos. 91–4001 through 91–4009). Initial checkout flights of each airframe took place at the company's plant in Marietta, GA. The airplanes were then delivered to the 411th Flight Test Squadron at Edwards AFB.

Assembly of Article 4008 began in 2000 and took 206,798 hours—9,627 more than originally planned. Delays in the construction of the EMD aircraft were due to design changes and modifications to the aircraft, parts shortages, and difficulty integrating hardware and software subsystems. As a consequence, the first flight of Article 4008 slipped from February 2001 to February 2002. Acceptance test flights were completed by April.[7]

On April 22, 2002, an Edwards AFB test pilot took off in Article 4008 from Dobbins Air Reserve Base, GA, for a ferry flight to Edwards AFB in conjunction with two F-15 chase aircraft and a KC-135R tanker. During rendezvous maneuvers shortly after takeoff, the F-22A ingested an 8.5-pound bird into the right engine. Despite extensive damage, the engine continued to operate normally and the pilot received no indication of a problem. The pilot aborted the flight, however, due to coincident but unrelated aircraft malfunctions. The bird-strike damage was discovered during postflight inspection.[8]

Following several weeks of repairs, Article 4008 was finally flown to Edwards AFB on May 31, 2002, becoming the last EMD aircraft delivered to the Air Force. There it served as the program's dedicated low-observables (stealth) test bed, weapons integration platform, and reliability and maintainability flight-test and evaluation aircraft. Article 4008 was carefully monitored to determine the resistance of the aircraft's stealth coatings to inclement weather as well as ordinary wear and tear commonly experienced by combat aircraft while being operated and maintained in field conditions.[9]

Weapons integration tests in 2006 included weapons-bay noise and vibration testing with the AIM-120D advanced medium-range air-to-air missile. In

---

7. Marvin E. Bonner, Edward Browning, Arthur Cobb, Travis Masters, Gary Middleton, Robert D. Murphy, Don M. Springman, and John Van Schaik, *Tactical Aircraft: F-22 Delays Indicate Initial Production Rates Should Be Lower To Reduce Risks*, GAO-02-298 (Washington, DC: United States General Accounting Office, March 2002).

8. *Executive Summary: Aircraft Accident Investigation, F-22A, S/N 91-4008, Dobbins ARB, Georgia, 22 April 2002* (U.S. Air Force, 2003).

9. Jeff Hollenbeck, "F-22 Raptor Team Delivers the Last Developmental Flight-Test Aircraft to USAF," F-16.net, *http://www.f-16.net/news_article1670.html*, June 3, 2003, accessed March 20, 2010.

February 2007, Article 4008 carried four 250-pound GBU-39 small-diameter bombs on board for structural load tests. During a similar test, while carrying eight of the weapons, the pilot performed a 360-degree negative-g roll. Incorrect engine trim settings resulted in a momentary dual flameout just before beginning the maneuver. The pilot was unaware of the problem because the engines automatically relit immediately, and the malfunction—which could have resulted in the loss of the aircraft—was only discovered through postflight data analysis.[10]

Separation testing began in September 2007 with the first subsonic bomb drop. Next, a series of high-speed release tests beginning at a speed of Mach 0.8 was followed by the first supersonic separation on February 11, 2008. Similar testing had to be conducted for each type of weapon carried by the F-22A.[11]

Cools

In September 2007, Lockheed Martin test pilot David P. "Cools" Cooley joined the F-22 Combined Test Force (CTF) at Edwards AFB. The 47-year-old Cooley had a great deal of experience in a wide variety of fighter aircraft. Born February 15, 1960, at Royal Air Force (RAF) Mildenhall, England, he grew up in Fairview Heights, IL, and attended Belleville East High School. In 1982 he graduated from the Air Force Academy in Colorado Springs, CO, with a bachelor of science degree in aeronautical engineering and a commission as a second lieutenant. He later earned a master of science degree in mechanical engineering from California State University at Fresno.[12]

Following flight training, Cooley was assigned to fly the F-111 and later served as an instructor pilot. He gained his first flight-test experience in 1989 with operational test and evaluation of weapons and systems for the F-111.

Cooley was selected to attend the Empire Test Pilot School in Wiltshire, England, as an Air Force exchange officer. Following graduation in 1992, he was assigned to the 445th Flight Test Squadron at Edwards AFB, where he conducted avionics testing and missile evaluations in the F-15. He also served as chief pilot for the U.S. Coast Guard RU-38 Twin Condor utility aircraft flight-test program.

---

10. Asif Shamim, "F-22 Flameout During SDB Flight Testing," F-16.net, *http://www.f-16.net/news_article2539.html*, October 1, 2007, accessed March 20, 2010.

11. SrA. Jason Hernandez, "Raptor Drops First Small Diameter Bomb," F-16.net, *http://www.f-16.net/news_article2528.html*; SrA. Julius Delos Reyes, "F-22 Raptor Performs First Supersonic SDB Drop," F-16.net, *http://www.f-16.net/news_article2975.html*, accessed March 20, 2010.

12. Jon Thurber, "David P. Cooley Dies at 49; Test Pilot Worked for Air Force, Lockheed Martin Before Fatal Crash," *Los Angeles Times* (March 30, 2009).

Lockheed Martin test pilot David P. "Cools" Cooley joined the F-22 Combined Test Force in September 2007. By late March 2009, he had accumulated a career total of more than 4,500 flight hours in a wide variety of aircraft, and colleagues considered him to be competent, thoughtful, meticulous, and well prepared. (U.S. Air Force)

In 1998, Cooley was assigned to the 410th Flight Test Squadron as operations officer and spent 2 years performing developmental testing of the F-117A. This included a variety of weapons integration tests. In 2000, Cooley became vice commandant of the U.S. Air Force Test Pilot School and was responsible for day-to-day operations of all school activities as well as mentoring students as a full-time flight instructor in T-38 and C-12 aircraft.[13]

After retiring from the Air Force in 2003, Cooley was hired by Lockheed Martin as F-117A chief test pilot. He spent the next 4 years conducting weapons separation and integration testing, developmental testing of avionics, and numerous classified test programs.[14]

Upon assignment to the F-22 CTF in 2007, Cooley accomplished F-22 transition training at Tyndall AFB, FL, over a 3-month period. After he completed 93 hours of academic instruction and 16 hours of simulator missions, his instructors commented on his outstanding knowledge of F-22 systems.[15]

Cooley's F-22 upgrade flight training began October 15, 2007, at Edwards AFB. Throughout a course of ground runs, basic handling sorties, and training in aerial refueling techniques and advanced handling characteristics, he demonstrated excellent knowledge and skills. Additionally, he was commended for his response to an aircraft emergency that was compounded by loss of radio communications during a spot evaluation check ride in November 2008.

By late March 2009, Cooley had accumulated a career total of more than 4,500 flight hours in a wide variety of aircraft, including over 1,000 hours of instructor time. He maintained dual qualification in the F-16 and F-22 and had 121 hours of time in the F-22 as a test pilot. Colleagues considered him competent, thoughtful, meticulous, and well prepared.[16]

High-G Test Maneuvers

Cooley was scheduled to fly a risk-reduction captive-carriage test of an air-to-air missile on March 25, 2009, to gather data on loads, flutter, vibroacoustics, and other effects on aircraft performance during high-g maneuvers with the left-side weapons bay door open. In preparation for the mission, he had practiced the flow of test maneuvers in the F-22 simulator to determine appropriate starting parameters for achieving the desired test points. The test conductor and

---

13. David P. Cooley, "Brief Summary of Flight Test Career," biographical information supplied to the Society of Experimental Test Pilots, May 2008.

14. Ibid.

15. Maj. Gen. David W. Eidsaune, *United States Air Force Accident Investigation Board Report: F-22A, T/N 91-4008* (U.S. Air Force, July 2009).

16. Ibid.

The Raptor 4008 is shown with the left-side weapon bay door open, exposing an AIM-9X missile. This is the same configuration that was flown in this aircraft on March 25, 2009, to gather data on loads, flutter, and vibroacoustic and thermal effects on aircraft performance during high-G maneuvers. (USAF)

lead loads engineer took notes during these simulations in order to incorporate lessons learned into the test cards that would be used during the mission. Additionally, two other F-22 test pilots practiced the same test maneuvers in the simulator and determined that the ideal starting points for the tests were an altitude of 25,000 feet above mean sea level (MSL) and airspeed of Mach 1.65. The test team recognized that the planned flight profiles required a significant amount of practice, concentration, and skill due to the physical and mental demands of the maneuvers.[17]

On the morning of the flight, Cooley briefed the test conductor, test director, chase pilot, range control officer, lead loads engineer, utilities and subsystem engineers, propulsion engineer, and a weapons integration expert on the details of the mission. He discussed how he intended to fly various test profiles, as well as safety procedures. Since the test points were to be conducted at very

---

17. Ibid.

high airspeeds and heavy g-loadings, the sortie was deemed a medium-risk test mission. In keeping with established protocols, the pilot, or anyone in the mission control, center could call "knock-it-off" if an unusual or dangerous situation developed.[18]

Following takeoff, the chase pilot performed a wet-dry check to verify that all panels and doors were secured and no fluids were leaking. He then executed two g-warmup turns to ensure that the aircraft and pilot's g-suit were operating properly and that the pilot was adequately prepared for the physical requirements of the high-g environment. G-warmup turns are commonly performed prior to high-g test maneuvers. Cooley also calibrated his instruments and verified that the weapon bay doors were functional before setting up for the first test point. The F-22 was equipped to transmit real-time voice communications and telemetry to engineers and test personnel in the mission control center so they could provide feedback to confirm that necessary data were captured during the execution of each test.

The first two test maneuvers involved subjecting the aircraft to a range of g-loads at specific airspeeds and altitudes with the left-side weapon bay door open and the missile launcher extended. The third test maneuver was to be identical to the first two except that it involved a different g-loading and the bay door was to be opened while the aircraft was subject to gravitational forces.

For each test run, Cooley was to roll inverted at Mach 1.65 and 25,000 feet MSL; perform a half split-S maneuver; then pull the throttles to idle, roll upright, and pull out of the dive. During the first run, Cooley opened the left-side weapon bay door while accelerating to target Mach number. He then rolled the aircraft and pulled straight down, achieving the required test parameters within seconds. Cooley sustained 6.8 g's, apparently recovering without difficulty. The only anomaly was the inadvertent opening of the infrared countermeasures-system doors, possibly when a pencil attached to the pilot's kneeboard struck a button on the control panel.[19]

Prior to the second test maneuver, engineers in mission control determined that the necessary g-load requirement was higher than had been the case for the previous maneuver. Cooley rolled the F-22 and paused briefly, letting the nose drop to 5 degrees nose low to ensure a starting speed of Mach 1.65. He then proceeded to pull into the spilt-S, sustaining 2 g's more than during the previous test. Cooley easily achieved the desired 8.8 g's—the most strenuous g-loading in the series—but as he executed his recovery procedure from 51 degrees nose low

---

18. Ibid.

19. Ibid.

During a 1999 test flight, Raptor 4002 is flown in an inverted dive with the left-side weapon bay open. This maneuver is similar to the one that was flown during the 2009 mishap in which the pilot was to roll inverted at Mach 1.65 and 25,000 feet MSL, perform a half split-S maneuver, then pull the throttles to idle, roll upright, and pull out of the dive. (U.S. Air Force)

at 22,400 feet MSL and Mach 1.6, he exclaimed, "Oh, man." Cooley received a brief respite as he rendezvoused with an airborne tanker for refueling.

With additional fuel on board, engineers determined a new desired g-loading for the final run. The new target was 7.8 g's, based on recalculation of the aircraft's actual gross weight. Cooley accelerated to Mach 1.65, rolled inverted, and began his dive. He opened the weapon bay door at 24,160 feet MSL, about 1,360 feet earlier than planned.

Cooley pulled the throttles to idle as per procedures but made only a slight lateral stick input. No pull was initiated that would help to start the turn around the split-S and thereby slow the ultimate loss of altitude. The aircraft remained inverted with the dive steepening from 65 to 72 degrees nose low. At 14,880 feet and Mach 1.49, with an 83-degree nose-low attitude, he finally rolled the aircraft upright and pulled full aft on the stick; however, the F-22 continued to descend rapidly. Cooley ejected just 3,900 feet above the ground while traveling at 765 knots equivalent airspeed. Unfortunately, this speed was 165 knots above the maximum ejection speed of his ACES II ejection system, and he sustained fatal injuries due to the blunt force trauma of the resultant windblast.[20]

20. Ibid.

A-LOC Incident

In the intensive investigation that followed, the physiological effects of high-g maneuvers were quickly singled out as the probable cause. Investigators found their first clue in the pilot's anti-g straining maneuver (AGSM), a technique aviators use to combat the effects of g-forces. Typically, to execute an effective AGSM, a pilot crisply inhales, holds, and then rapidly exhales air at 3-second intervals. Based on evaluation of audio recordings from the F-22 accident, however, Cooley's AGSM performance was ineffective. His breathing was described as "labored and strained," with "long grunting air exchanges at 5- to 6-second intervals."[21]

Based on his flight experience, Cooley should have been absolutely proficient in performing the AGSM to mitigate the effects of high g-forces during flight. His most recent AGSM evaluation and training in January 2007, however, had resulted in a rating of "average." This means that his AGSM performance "had not been mastered fully" and that minor AGSM performance errors impacted his breathing and muscle-straining technique. Additionally, an audio recording of an earlier high-g test mission on March 23, made just days prior to the mishap flight, indicated that he was using improper AGSM breathing technique.

During the second test run on the day of the mishap, Cooley's AGSM seemed particularly labored. During his third run, he did not execute a proper AGSM. He was not heard to inhale, but instead made labored grunts and groans accompanied by nearly continuous exhalation. As a result he became susceptible to A-LOC. The aircraft itself was functioning normally, and no design or airworthiness problems were identified.

Investigators determined that "there was a progressive breakdown in [the pilot's] AGSM technique during each successive test maneuver." As a result of his physical impairment due to A-LOC, Cooley lost situational awareness while the aircraft was in a steep dive. He failed to recover, ejected at high speed, and died due to blunt-force trauma from air blast during ejection. The Accident Investigation Board concluded that the cause of the mishap was the pilot's "adverse physiological reaction to high acceleration forces, resulting in channelized attention and loss of situational awareness."[22]

G Physiology

The physiological effects of elevated g-forces on the human body have been studied for many years, resulting in in-depth understanding of the topic as

---

21. Ibid.

22. Ibid.

well as the development of various g countermeasures. The reader is referred to the many sources available on the topic, and this text will summarize only the key concepts.

The essential concept in understanding g physiology is that of centrifugal force. This is the vector of acceleration that pushes from the center of a circle to its outer perimeter during rotation about the axis. Because of the aerodynamics of flight, this almost always translates to the vertical axis of the body (pushing towards the feet) and, by conventional nomenclature, is called the $+G_z$ axis (as opposed to the positive and negative G_x and G_y transverse and lateral axes, respectively). The unit used in discussing acceleration forces is the g, which is defined as the normal pull of gravity (32 feet/second^2 or 9.81 meters/second^2). Thus, 2 g's would be double that pull, or twice the force of gravity; 3 g's would be three times the force of gravity, and so on.

The g's experienced in flight primarily affect the cardiovascular and the pulmonary systems of the body. These effects result in neurologic consequences that can range from grayout to the full G-LOC. The concept of A-LOC, as described in the F-22 accident report, is a relatively new concept in the field of g physiology and will be discussed later in greater depth. All references to g's in this chapter refer to $+G_z$ for brevity.

Cardiovascular System

As g's on the pilot increase during flight, it becomes more difficult for the heart to pump the blood upward, against the g-force, into the pilot's head. Blood tends to be pulled from the upper portions of the pilot's body, resulting in reduced blood flow to the brain, and pooling of blood in the buttocks and lower extremities. A key factor in this situation is the distance between the heart and the brain, or the heart-to-head distance, since the weight of the column of blood in this zone increases linearly with the amount of g's applied. The workload on the heart, and its resulting oxygen demand, increases exponentially with the increasing force against which it must pump. This force is known as afterload. Moreover, as blood flow to the aortic arch and head region decreases, specialized nerve cells in the carotid arteries detect an associated drop in blood pressure. These baroreceptors trigger a compensatory reflex response to increase heart rate, increasing cardiac output in order to maintain blood flow to the brain. Vascular systems, or beds, are also recruited to increase blood pressure through constriction if there is time for the physiological response to engage.

With higher g-forces reducing blood flow to the brain, the resultant heart rate increase and vasoconstriction significantly increase cardiac workload and thus myocardial oxygen demand. Simultaneously, the g's reduce venous return from the pilot's lower extremities, resulting in pooling of blood in the lower extremities, thus reducing the amount of blood returning to the heart. This pooling results in a reduction in the amount of blood pumped with each heart contraction (i.e., decreased stroke volume). Ultimately, with increasing or sustained g's, the cardiovascular response will not be adequate for maintaining blood flow sufficient for the pilot to retain consciousness. The result is G-LOC.[23]

Pulmonary System

G-forces also affect lung function and oxygenation of blood cells. The key concept involves matching the ventilation of the lung alveoli with the perfusion of pumped blood such that blood cells are properly oxygenated with the respired air. In this context, physiologists normally describe three zones in the lung. At the top portion (Zone 1), there is no perfusion of blood through the capillaries surrounding the alveoli, due to the effect of gravity on the hydrostatic column of blood. Even though this zone is ventilated with air breathed in from the atmosphere, there is no blood flow and, thus, no oxygenation of blood. Lower down in the lung is the zone (Zone 2) where the alveoli are both ventilated with air and perfused with blood. This combination is necessary for effective oxygenation of the blood and subsequent delivery of oxygenated blood to the body tissues. Hence, this is the most physiologically relevant zone. Lower down in the lung is Zone 3, where the blood settles due to the relative differences in hydrostatic pressure between blood and air. As a result, the alveoli in these areas tend to be compressed to the point of collapse. This zone is therefore perfused, but not ventilated. With normal activity in a 1-g environment, the percentage of Zone 1 and Zone 3 relative to the physiologically effective Zone 2 in terms of lung function is very small. In other words, ventilation and perfusion are generally well matched.[24]

However, this situation changes when g's are applied. The effect might be illustrated by soaking a sponge in water, putting it in a bucket, and then spinning the bucket in a large circle. Much of the water would be pushed by the centrifugal force from the top part of the sponge to the bottom.

As g's increase, blood is increasingly pulled to the lower portion of the lung, creating a vertical perfusion gradient. As the hydrostatic pressure—and blood flow—drops in the upper portions of the lungs, there is less perfusion in the upper regions. In effect, the top zone of the lung (ventilated but not perfused) increases. Meanwhile, due to the increased hydrostatic weight of the blood

---

23. Robert D. Banks, James W. Brinkley, Richard Allnutt, and Richard M. Harding, "Human Response to Acceleration," chapter 4 in *Fundamentals of Aerospace Medicine*, ed. Jeffrey R. Davis, Robert Johnson, Jan Stepanek, and Jennifer A. Fogarty, 4th ed. (Philadelphia: Lippincott Williams & Wilkins, 2008), pp. 83–109.

24. Ibid.

due to g's, there is increased blood flow to the lower portions of the lungs, and as a result there are a greater number of alveoli that collapse because they are engorged with blood. Thus, the bottom zone of the lung (perfused but not ventilated) increases due to increased g's.

Therefore, the overall effect of g's on the lung causes both the top zone (Zone 1) and the bottom zone (Zone 3) to increase at the expense of the physiologically useful middle zone (Zone 2). The end result of this is that approximately 50 percent of the blood that passes through the lungs does not achieve a gas exchange with the alveoli, and therefore this blood is not oxygenated.[25]

Visual System: Grayout and Blackout

As the blood is pumped to the brain, the first artery to branch off of the internal carotid artery is the ophthalmic artery, which goes to the eye. Hence, as g's are applied in the head-to-toe direction, blood flow through this artery decreases. Intra-ocular pressure may be another factor in reducing blood flow to the retinal tissue as blood pressure to the eye decreases. In any case, the retina is made of specialized nervous tissue and, like the brain, is particularly sensitive to oxygen depletion. During early stages of g-onset, particularly when the onset is gradual, peripheral vision is lost due to the retinal depletion of oxygen, and there may be an overall dimming of the visual image. This condition is called grayout, or tunneling, and is considered one of the first signs that a pilot has impaired blood flow to the head region. Interestingly, the executive functions of the pilot are located in the brain's cerebral cortex, which is even higher, and so farther away from the heart. Thus, vision grayout could be a good warning for relatively impaired cerebral blood flow since both are nervous-system tissues that are very sensitive to blood flow reduction. However, since these areas are served by different arterial pathways, the heart-to-tissue height is not the only factor.

As the duration or magnitude (or both) of the g's increases, retinal blood supply is further diminished, resulting in loss of central vision without loss of consciousness. This condition is known as blackout. Because blackout is technically not a loss of consciousness but a loss only of vision, the aviator will generally still have his or her mental and motor faculties available. Ultimately, however, if g levels are maintained or increased, oxygenation will fall below the level required to maintain consciousness, potentially resulting in G-LOC.[26]

---

25. Ibid.
26. Ibid.

Stoll Curve

Based on data gathered on healthy individuals in centrifuge studies, in the 1950s Alice Stoll developed the G Time-Tolerance curve (also known as the Stoll curve) to indicate the general progression from normal mental function to grayout, blackout, and G-LOC. An individual generally can tolerate a very high amount of gravitational force for a few seconds, due to the oxygen reserves in the brain tissue. Beyond this, time until grayout, blackout, and G-LOC becomes a function of g's pulled and individual tolerance factors. Without some level of g-protection, the average individual will not tolerate greater than 5 g's for more than 5 seconds without loss of consciousness. If the g onset rate is more gradual, then the pilot experiences the visual symptoms of grayout and blackout. Of course, if the g's are of sufficient magnitude

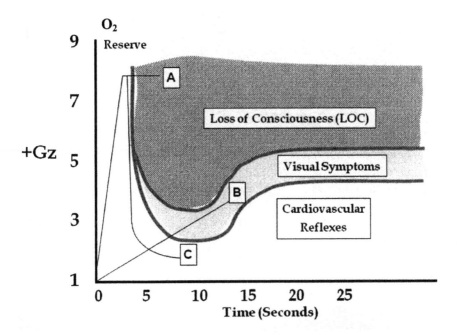

The G Time-Tolerance curve (Stoll curve) illustrates the effects of high g-forces in a general progression from normal mental function to grayout, blackout, and ultimately g-induced loss of consciousness. An individual can generally tolerate a high g level for a few seconds, due to oxygen reserves in the brain tissue. Without some type of g-protection, the average individual will not tolerate greater than 5 g's for more than 5 seconds without a loss of consciousness (the region denoted by box A). If the g-onset rate is more gradual, the pilot experiences the visual symptoms of grayout and blackout (the region denoted by box B). If g levels are sufficiently low, then no symptoms will occur (box C). The dip in this region of tolerance is the result of the time delay required for the cardiovascular baroreceptor reflexes to manifest. (Author's collection)

for sufficient duration, G-LOC will ensue. However, if the magnitude or duration is lower, then the cardiovascular responses (autonomic constriction of blood vessels, increased heart rate, and the like) will enable the pilot to maintain consciousness, with or without the continued presence of visual symptoms. Finally, if the g's are sufficiently low, then no symptoms will occur. The dip in this region of tolerance is the result of the time delay required for the cardiovascular baroreceptor reflexes to manifest. On the other hand, it is certainly possible to proceed directly from consciousness to unconsciousness under certain conditions of rapid g-onset rate and magnitude.[27]

Since the Stoll curve was based on centrifuge data from healthy volunteers in the 1950s, it is merely a generalization. There are many variables (individual differences, dehydration, previous recent g exposure, general health, blood pressure, etc.) that affect an individual's physical responses to g-forces. Only guarded predictions can be made for a specific individual in an actual operational setting. More recent attempts have been made to model the effects of g using more recent data and more sophisticated computer-modeling techniques. But the classic Stoll curve nevertheless demonstrates in an elementary fashion the human physiologic response to g stress.[28]

Incapacitation Due to G-LOC

After onset of G-LOC, there is a subsequent period of incapacitation that can range from approximately 15 seconds to over 1 minute. The overall period of incapacitation is usually considered to involve two phases: absolute incapacitation and relative incapacitation. Absolute incapacitation (AI) is the period of time during which the individual is actually unconscious. Based on centrifuge studies, the average AI time is 12 seconds, with a range of 9 to 22 seconds. Relative incapacitation (RI) is the recovery period of time following the return of consciousness before a pilot is able to regain situational awareness and control of the aircraft and related tasks. This period also averages 12 seconds, with a range from 5 to 40 seconds. Symptoms of the RI phase include disorientation, confusion, stupor, and apathy. Additional symptoms that may be present are event amnesia, convulsive flailing, tingling, euphoria, anxiety, and nausea.[29]

---

27. Alice M. Stoll, "Human Tolerance of Positive G as Determined by the Physiological End Points," *Journal of Aviation Medicine* 27 (1956): 356–367.

28. Thomas W. Moore, Dov Jaron, Leonid Hrebien, and David Bender, "A Mathematical Model of G Time-Tolerance," *Aviation, Space, and Environmental Medicine* 64 (1993): 947–951.

29. Banks et al., "Human Response to Acceleration."

Characteristics of A-LOC

The U.S. Navy introduced the term "almost-loss of consciousness" in the late 1980s, and it is now a recognized phenomenon that can occur at g levels insufficient to cause G-LOC.

With A-LOC, aviators do not completely lose consciousness. Rather, they experience a wide variety of cognitive, physical, emotional, and physiological symptoms that can cause significant sensory, motor, and cognitive impairment. Depending on the circumstances, this phenomenon can lead to the loss of aircraft, as well as loss of life, as is evidenced by the F-22A mishap.[30]

In a carefully designed and controlled centrifuge study, researchers discovered 66 episodes of A-LOC during 161 +G_z exposures in specially instrumented research subjects. There were many episodes of a variety of symptoms to include amnesia, confusion, euphoria, difficulty in forming words, and reduced auditory acuity, which occurred in multiple subjects at varying +G_z levels. Of particular note, one of the most common symptoms of the A-LOC syndrome was a disconnection between cognition of a situational event and the ability to act upon it. These episodes of A-LOC were associated with particular patterns of decreased oxygenation compared to those +G_z exposures in which the individuals did not experience A-LOC symptoms.[31]

A-LOC is a phenomenon distinct from either grayout or blackout because there appears to be evidence of actual cerebral involvement rather than ocular activity only. There is apparently a loss of blood flow and/or oxygenation to various parts of the brain resulting in symptoms short of total loss of consciousness. In this sense, A-LOC is more similar to, yet still distinct from, the period of relative incapacitation experienced by pilots who have undergone G-LOC, since the A-LOC syndrome is not specifically associated with a total loss of consciousness. There are thought to be effects on situational awareness, spatial disorientation, altered cognitive function, and states of awareness.[32] One survey of operational U.S. fighter pilots from the Air Force, Navy, and Marines reported that 14 percent had operationally experienced A-LOC symptoms at least once.[33]

---

30. Barry S. Shender, Estrella M. Forster, Leonid Hrebien, Han Chool Ryoo, and Joseph P. Cammarota, Jr., "Acceleration-Induced Near-Loss of Consciousness: The 'A-LOC' Syndrome," *Aviation, Space, and Environmental Medicine* 74, no. 10 (2003): 1021–1028.

31. Ibid.

32. Sinha and Tyagi, "Almost Loss of Consciousness."

33. K.L. Morrissette and D.G. McGowan, "Further Support for the Concept of a G-LOC Syndrome: A Survey of Military High-Performance Aviators," *Aviation, Space, and Environmental Medicine* 71 (2000): 496–500.

A survey of Royal Australian Air Force pilots found that 52 percent had experienced A-LOC symptoms, to include abnormal sensation in the limbs, disorientation, and confusion. A-LOC syndrome is obviously not addressed in such things as the Stoll curve, further underscoring the need for some degree of revision of previous models of g physiology.[34]

There are several techniques and pieces of equipment available for mitigating the incapacitating forces that lead to G-LOC.

The Anti-G Straining Maneuver (AGSM)

The AGSM is an all-encompassing term for what has historically been known as the M-1, L-1, or Hook maneuver. Though there are subtle differences in the way each of these techniques is performed, the two basic components they both utilize to affect g protection are isometric muscle contraction and forced exhalation.

Research and operational experience have shown that forced isometric contraction of the muscles in the legs and abdomen greatly increases tolerance to g's. This is a result of the increased venous return to the heart, with resultant increased blood flow to the brain. The forced exhalation component of the maneuver can be described as a forced exhalation against a closed glottis. When properly performed, the pilot makes a sound similar to saying the word "hook," hence the name of the maneuver. When this is performed with a hard diaphragmatic contraction, the intrathoracic chest pressure increases, which in turn increases the blood pressure to the head. It is also possible that the increased pressure more evenly distributes the perfusion gradient in the lung against the g's, thereby lessening the ventilation-to-perfusion mismatch seen with increased g. The same increase in chest pressure, however, will also impair venous return to the heart. For this reason the chest pressure must be briefly released on an intermittent basis to allow for venous return. This release is made through a small, partial air exchange, followed immediately by the next diaphragmatic contraction. Research has shown that the ideal time to hold the breath before the short release is 3 to 5 seconds. If held longer, venous return becomes problematic; if done more frequently, fatigue becomes a factor. Generally, a well-performed AGSM will yield up to 4 g's of additional

---

34. Caroline A. Rickards and David G. Newman, "G-Induced Visual and Cognitive Disturbances in a Survey of 65 Operational Fighter Pilots," *Aviation and Space Environmental Medicine* 76, no. 5 (2005): 496–500.

protection. The limiting factor is muscle fatigue and focus on correct execution of the straining maneuver.[35]

G-Suit

The g-suit is a pair of pants, worn over the flight suit, that contains a series of interconnected air bladders located at the abdomen, thighs, and calves. These bladders are connected to the aircraft with a hose and a valve fitting. As g's increase, the valve automatically opens and forces air into the bladders, which in turn expand. The expansion of the bladders compresses the tissues of the lower body, thereby reducing the pooling of blood in this region. The result is an increase in available blood volume to be pumped by the heart, with a resultant increase in blood flow to the brain. Although older g-suit technology (such as the CSU-13B/P still worn in the F-15 and F-16 fighter aircraft) provides less physical coverage, it allows the pilot to withstand approximately 1.5 g's more than can be withstood with new, advanced g-suit technology used in the F-22 Raptor and other advanced fighters.

Positive-Pressure Breathing (COMBAT EDGE)

The most recent advance in combating A-LOC and G-LOC involves positive-pressure breathing and is called the Combined Advanced Technology Enhancement Design G Ensemble (COMBAT EDGE). It combines positive-pressure breathing with a chest counter-pressure vest and a full-body coverage g-suit, all linked to the g-valve connecting the suit to the aircraft. The positive-pressure breathing, which increases linearly starting at 4 g's and continues through 60 millimeters of mercury pressure at 9 g's, is continuous as long as the aircraft is experiencing gravitational force. The positive-pressure breathing leads to an increased intrathoracic pressure, increasing cardiac output during increased g's. The counter-pressure jerkin vest is a safety measure that prevents lung over-inflation injuries. Many operational units currently fly without the vest because they consider it too cumbersome for the protection it affords. But with advanced g-suit full-coverage technology, less effort must generally be expended on performing the AGSM, so aircrew are much less fatigued and are thus able to sustain prolonged or recurrent exposures to high g for longer periods of time. COMBAT EDGE is not necessarily designed to increase the maximum g attainable (although this may be a result) but, rather, to reduce the fatigue associated with the AGSM, thereby increasing endurance. The

---

35. Larry P. Krock, "Aeromedical Issues Related to Positive Pressure Breathing for $+G_z$ Protection" (Brooks Air Force Base, TX: Armstrong Laboratory Crew Systems Directorate TP-1992-0024, 1993).

pilot can keep talking and breathing throughout high g's, though exhalation and communication require effort and some practice in order to be executed with proficiency.[36]

In a recent survey, F-22 pilots reported getting, on average, about 1.5 g's more protection from advanced full-coverage technology anti-g suits, with much less fatigue for a given g, and highly recommend that the equipment be retrofitted into legacy fighter aircraft.[37]

Factors in the F-22A Mishap

In the wake of the F-22A accident, investigators determined that the pilot was wearing an Advanced Tactical Anti-G Suit (ATAGS), a newer and improved version of a conventional g-suit. He was likewise utilizing the COMBAT EDGE positive-pressure breathing device, along with its anti-g vest. All of this equipment was properly inspected prior to the mishap mission, and post-mishap operability tests confirmed that all components functioned as designed during the test mission.

Of greater note, however, was the pilot's AGSM. The test mission consisted of three test maneuvers with high g exposures. Based on the g's encountered, this mission qualified as a physically demanding mission. The pilot sustained high g for an average of 15 seconds during each test and subsequent recovery period. However, there was a progressive breakdown in his AGSM technique during each successive test maneuver.[38]

Analysis of the audio recording of the mission by an aerospace physiologist revealed that the pilot's AGSM technique was not fully effective in terms of classical AGSM. The pilot may have had an acceptable level of g protection, based on the fact that he successfully flew the first two profiles, but level of protection in the third appears suspect.

During the first two test maneuvers, the AGSM was marked by long grunting air exchanges at 5- to 6-second intervals. As noted, the ideal time between intervals is 3 to 5 seconds. On the third test maneuver, no audible inhalation is discernible; rather, the pilot can be heard continuously grunting his expirations, which constitutes definite evidence of an improperly performed AGSM. This improper technique may have compromised intrathoracic pressure, possibly resulting in compromised blood flow to the head. Also, contrary to what was expected at this point in the flight profile, there was no aggressive roll to an

---

36. Ibid.

37. Ibid.

38. Eidsaune, *United States Air Force Accident Investigation Board Report: F-22A, T/N 91-4008.*

upright position. Thus, the aircraft passed through the test band and continued to descend in a fast, steep dive.

It is possible that the pilot was suffering from the physiological effects of A-LOC. He did not experience G-LOC because throughout the mishap test maneuver he continued to command the aircraft (based on the presence of stick inputs) and made a distressed statement just prior to ejection. He was therefore conscious but impaired for some reason in terms of his flight performance, possibly by acceleration forces. His attentional resources may have been challenged for a variety of reasons. As a result, he lost good situational awareness with regard to the aircraft's altitude, airspeed, and dive angle. He eventually recognized the unsafe attitude, altitude, and airspeed of the aircraft and initiated ejection. Unfortunately, this ejection was outside the limits of the ejection seat's design envelope.[39]

Cooley's most recent centrifuge evaluation and training had been on January 31, 2007, more than 2 years prior to the mishap, which is not an unusual or problematic amount of time. During this training and evaluation, with regard to the AGSM, he was rated average. An average rating is defined as "AGSM performance had not been mastered fully and minor AGSM performance errors impact AGSM technique." But average is deemed acceptable, and he was certified for high-g flight.[40]

Successive +G_z Exposures

The question of whether the successive g exposures may have led to fatigue on the part of the pilot remains unresolved. If fatigue had been a factor, g-tolerance would wane with successive exposures, accounting for the ineffective AGSM during the third test run. Researchers not associated with the mishap investigation recently published a study showing the physiologic response to successive +G_z exposures in simulated air combat maneuvers (SACMs). What the researchers found was an average increase in calculated +G_z tolerance of approximately 0.35 g following the first g exposure. In fact, they found that simulated air combat maneuvers with short g pauses produced greater increase in g-tolerance than SACMs with long g pauses. The authors attributed the increased g-tolerance to a carryover of cardiovascular compensatory responses, primarily vasoconstriction, "with possible contribution from greater venous return and baroreflex enhancement." It is therefore reasonable to conclude

---

39. Ibid.

40. Ibid.

that Cooley's successive g exposures did not lead to loss of g-tolerance in the mishap, unless fatigue in general was an overriding factor.[41]

G-LOC and Age

Another factor to consider is pilot age. Cooley was 49 years old. The question of whether g-tolerance decreases with age due to the inevitable effects of aging on human physiology had been addressed years earlier during research in which 53 healthy Air Force crewmembers ranging in age from 26 to 55 years old took part in a centrifuge study designed to determine the effects of age on resting $+G_z$ tolerance. The results showed that, among characteristics studied, only age was positively correlated with tolerance of rapid-onset g-forces. The study authors concluded that aging might, in fact, offer some protection from g-stress. In any case, there is no evidence that aging led to a decrease in g-tolerance. Therefore, it may reasonably be concluded that age was not a factor in the loss of g-tolerance in the F-22A mishap.[42]

G-Stress Remains a Threat

In light of numerous advances in modern aerospace technology, it is tempting to regard the problem of g-tolerance in high-performance aircraft as having been solved. Maximum g limits can be designed into the FCSes of modern aircraft; pilots can be trained to perform the AGSM; the g-suit has been refined over the years to be more effective; and positive-pressure breathing with a counter-pressure vest may be utilized. Indeed, with modern radar, long-range missiles, and global positioning technology, some critics argue that the days of dog fighting between combat aircraft are a thing of the past; thus, g-stress should no longer be a concern. The F-22A mishap, however, vividly illustrates that the problem of g-stress remains an issue to be dealt with.

One final point to consider is that of the flight-test profile itself. Had the third maneuver been begun at a higher altitude, Cooley would not have had to pull as hard to make the required turns (which generated the attendant high g's). This was a supervisory and test-management plan issue in a general sense, and it could have been addressed by building greater leeway into the final test profile. Lastly, the pilot should have realized that at the altitudes and airspeeds he was flying for the inverted test, there was little room for error (g, flying, or physiologic); any delay that occurred in pulling while inverted and rolling out would contribute to Cooley's inevitable encounter with the

41. Lalande and Buick, "Physiologic $+G_z$ Tolerance Responses," pp. 1032–1038.

42. D.H. Hull, R.A. Wolthuis, K.K. Gillingham, and J.H. Triebwasser, "Relaxed $+G_z$ Tolerance in Healthy Men: Effect of Age," *Journal of Applied Physiology* 45, no. 4 (1978): 626–629.

dangerous circumstances that caused the mishap. Many factors—situational awareness, perhaps some spatial disorientation, altered/impaired states of consciousness (possible A-LOC), and the lack of good checks on the profile being flown—contributed to the loss of life and aircraft. Ideally, the lessons learned will be remembered and thoughtfully considered for future flight tests at high g.

Part 3: Organizational Factors

The XB-70A Valkyrie was a Mach 3–capable bomber prototype, of which only two were built. On June 8, 1966, the second XB-70A was flown in formation with several other aircraft in order to take publicity photos for engine manufacturer General Electric. (USAF)

CHAPTER 7

Decision Chain Leading to the XB-70/F-104 Midair Collision

Many people with varying levels of responsibility are involved in the conduct of any aircraft operation. Administrators, mission planners, maintenance crews, and flightcrew members each play a role and make decisions that affect the mission's outcome. The XB-70 accident is a story of a chain of decisions that led to tragedy. To use the analogy of James Reason's "Swiss cheese" model, there were numerous opportunities to "close the holes" in the layers of latent conditions leading to this mishap.[1]

World's Largest Research Aircraft

The B-70 Valkyrie was designed as a strategic bomber capable of attaining Mach 3 speeds and delivering nuclear or conventional weapons. After a succession of policy changes led to the cancellation of the bomber program, two prototypes—designated XB-70—capable of flight at 2,000 miles per hour and altitudes of about 70,000 feet served as platforms to collect flight-test data for use in the design of future military and civilian supersonic aircraft.

A design study in January 1954 recommended the development of a long-range, high-performance bomber with a high-speed, high-altitude supersonic dash capability. By March 1957, however, the aircraft's specifications called for a bomber capable of cruising at Mach 3 speeds for an entire mission, as opposed to a subsonic cruise/supersonic dash mission profile. First flight of the prototype was expected in December 1961, and the first wing of Air Force Valkyries was to be operational by August 1964.[2]

In the fall of 1958, however, funding limitations were causing schedule delays. Additionally, President Dwight Eisenhower had begun to doubt the

---

1. James T. Reason, *Managing the Risks of Organizational Accidents* (Aldershot, U.K.: Ashgate Publishing Company, 1997).
2. Jeannette Remak and Joe Ventolo, Jr., *XB-70 Valkyrie: The Ride to Valhalla* (St. Paul, MN: MBI Publishing Company, 1998).

need for the B-70 program, concluding that the bomber made very little military sense—especially in view of intercontinental ballistic missiles just entering service. At the same time, there was growing interest in an American supersonic transport (SST) with commercial applications; NASA had several SST studies under way that would benefit from data acquired by the XB-70 test program.

After John F. Kennedy succeeded Eisenhower, he found that a feared United States–Soviet "missile gap" did not actually exist and that Soviet capabilities had been grossly exaggerated during the heat of the presidential campaign. In March 1961 Kennedy directed that the B-70 program be reoriented toward research and development. With the exception of three XB-70 prototypes that were to be built, one of which was never completed, plans for B-70 production were terminated.[3]

The XB-70 configuration included a delta-winged planform with a long forward fuselage and canards. It was powered by six General Electric (GE) YJ93 afterburning turbojet engines, each providing up to 30,000 pounds of thrust. The gross weight was around 500,000 pounds. To achieve Mach 3 performance, the XB-70 was designed to ride its own shock wave, much as a surfer rides an ocean wave. For this wave-rider concept, the outer wing panels were hinged. During takeoff, landing, and subsonic flight, they remained in the horizontal position to increase lift and improve lift-to-drag ratio. During supersonic cruise, the wing panels were lowered, reducing drag and improving directional stability at high Mach numbers. With a length of 189 feet and a span of 105 feet, the XB-70 was the world's largest research aircraft.

The first XB-70 made its maiden flight from North American's Palmdale facility to Edwards AFB on September 21, 1964. Over the next 2 years, contractor and Air Force pilots conducted airworthiness and performance tests. Although intended to cruise at Mach 3, the first XB-70 was found to have poor directional stability above Mach 2.5 and made only a single Mach 3 flight. Despite these problems, the early flights provided data on several issues facing SST designers, including aircraft noise; operational problems; control system design; comparison of wind tunnel predictions with actual flight data; and high-altitude, clear-air turbulence.[4]

NASA wind tunnel studies led engineers at North American to build the second XB-70 (Air Force serial no. 62–0207) with an additional 5 degrees of dihedral on the wings. This aircraft, Air Vehicle Two (AV-2), made its first flight on July 17, 1965, with North American test pilot Al White and Col. Joseph

---

3. Dennis R. Jenkins and Tony R. Landis, *Valkyrie: North American's Mach 3 Superbomber* (North Branch, MN: Specialty Press, 2004).

4. Ibid.

F. Cotton at the controls. The same crew took AV-2 up to Mach 3.05 at 72,000 feet on January 3, 1966, during its 17th flight. The design changes resulted in greatly improved handling qualities, and the airplane maintained full speed for 32 minutes, on a course covering eight states. In March 1966 White and Van Shepard took AV-2 up to 74,000 feet, the highest-altitude flight of the entire XB-70 program. The fastest speed achieved during the program, also in AV-2, was Mach 3.08 on April 12, 1966. By early June 1966, 45 flights had been completed with AV-2, including 9 Mach 3 cruise demonstrations.

At the same time, NASA and Air Force officials signed a joint agreement that would allow the second prototype to be used for high-speed research flights in support of the SST program. AV-2 was selected because its aerodynamics, inlet controls, and instrument package were superior to those of the first aircraft. NASA research flights made to evaluate typical SST flight profiles and study the problems of sonic booms on overland flights were to begin in mid-June, following the completion of Phase I contractor tests of the vehicle's

airworthiness. NASA research pilot Joseph A. Walker was selected as the project pilot. He began preparing for his role by flying chase during contractor and Air Force test flights.[5]

Chief Research Pilot

Walker was born February 20, 1921, in Washington, PA, where he lived until graduating from Washington and Jefferson College in 1942 with a bachelor of science degree in physics. During World War II he joined the Army Air Forces and, following flight training, flew P-38 fighters and F-5A reconnaissance aircraft in North Africa. For his wartime service he earned the Distinguished Flying Cross and the Air Medal with seven oak leaf clusters.

In March 1945, Walker left military service and found work as a physicist with the National Advisory Committee

NASA research pilot Joseph A. Walker was assigned to the XB-70 test program in 1965. To prepare for flights in the bomber, he was tasked to fly chase as a safety observer, giving him the opportunity to study the Valkyrie in flight and become familiar with some of its characteristics. (NASA)

---

5. Peter W. Merlin and Tony Moore, *X-Plane Crashes—Exploring Experimental, Rocket Plane and Spycraft Incidents, Accidents and Crash Sites* (North Branch, MN: Specialty Press, 2008).

for Aeronautics (NACA) at the Lewis Propulsion Research Laboratory (now NASA Glenn Research Center) in Cleveland, OH. There he eventually became a research pilot and conducted aircraft icing research.

Walker transferred to the NACA High-Speed Flight Research Station (now Dryden) at Edwards AFB in 1951. There he piloted numerous jet-powered and rocket-powered research aircraft, including the D-558-1, D-558-2, X-1, X-1A, X-1E, X-3, X-4, X-5, and X-15. He also flew programs involving the modified and unmodified production variants of the F-100, F-101, F-102, and F-104, as well as the B-47. Using a highly modified F-104 and flying a zoom-climb profile to 90,000 feet, he participated in pioneering work to develop reaction controls for high-altitude flight above the sensible atmosphere. In 1955 he was promoted to chief research pilot.[6]

Walker had extensive experience in all aspects of flight research and was well respected by his peers. Fellow research pilot Milton Thompson described him as a demanding boss with a quick temper but emphasized that Walker combined experience with sound engineering judgment.[7]

Walker received the NACA Exceptional Service Medal in 1955 for his actions during an in-flight emergency involving the X-1A and its B-29 mother ship. He had piloted the first NACA-sponsored flight of the X-1A on July 20, 1955. During preparations for a second launch on August 8, an explosion shook the X-1A while it was still mated to the belly of the B-29. Walker calmly shut off all power switches, vented the cockpit pressurization, and climbed into the B-29 with assistance from other crewmembers.

Although it had been relatively small, the blast caused the rocket plane to drop a few inches and its landing gear to extend. This made it impossible to safely land the B-29 with the X-1A still attached. After Walker and the other crewmen determined that it would be impossible to raise the rocket plane's landing gear or safely jettison its propellants, the X-1A was jettisoned over the Edwards AFB bombing range. Walker and the B-29 crew were later recognized for "outstanding bravery beyond the call of duty" during their valiant, but unsuccessful, efforts to try to save the X-1A.[8]

In 1958, Walker was selected for the Air Force–sponsored Man-in-Space Soonest project, the goal of which was to place a human in orbit by October 1960. Following the creation of NASA later that year, however, the

---

6. Biographical Information Files, DFRC Historical Reference Collection, NASA Dryden Flight Research Center.

7. Milton O. Thompson, *At the Edge of Space: The X-15 Flight Program* (Washington, DC: Smithsonian Institution Press, 1992).

8. Merlin and Moore, *X-Plane Crashes.*

new civilian space agency was given responsibility for crewed space exploration programs, which came to include those of the Mercury ballistic capsule and the X-15 rocket plane.[9]

In 1960, Walker became the first NASA pilot to fly the X-15, following the contractor checkout flights by Scott Crossfield. He eventually flew the research aircraft 25 times, achieving two flights above 50 miles altitude, which qualified him as an astronaut. The first, in January 1963, reached 271,000 feet. Several months later he flew the highest X-15 flight, attaining an altitude of 354,200 feet. He also flew the fastest flight in the basic configuration, achieving a speed of 4,104 miles per hour. In recognition of his achievements, President Kennedy presented him with the 1961 Collier Trophy.[10]

Walker received many other awards and honors during his 21 years as a research pilot, including the 1961 Harmon International Trophy for Aviators, 1961 Kincheloe Award, and 1961 Octave Chanute Award. He received an honorary doctor of aeronautical sciences degree from his alma mater in June 1962 and was named Pilot of the Year in 1963 by the National Pilots Association. He was a charter member of the Society of Experimental Test Pilots and one of the first to be designated a fellow.[11]

In October 1964, Walker became the first to pilot NASA's Lunar Landing Research Vehicle (LLRV), used to develop piloting and operational techniques for the Apollo Moon landings. He piloted 35 flights in the LLRV, including a special demonstration for visitors during which he had to make a critical decision in order to avoid disaster. Walker was to take off, demonstrate lunar simulation mode, hover, and land on the aircraft parking apron in front of the NASA hangars. As he started his approach, it was obvious that he would overshoot the landing site. Recognizing his problem, Walker switched the LLRV into standard vertical flight mode, using the jet engine to arrest horizontal motion. He then brought the vehicle to an uneventful landing. Walker left the LLRV project in 1965 when he was assigned to fly the XB-70.[12]

Before flying the XB-70, Walker was assigned to fly chase as a safety observer. This gave him the opportunity to study the Valkyrie in flight and become familiar with some of its characteristics. For eight of these missions, Walker piloted

---

9. Loyd S. Swenson, Jr., James M. Grimwood, and Charles C. Alexander, *This New Ocean: A History of Project Mercury* (Washington, DC: NASA SP-4201, 1998).

10. Thompson, *At the Edge of Space.*

11. Biographical Information Files, DFRC Historical Reference Collection.

12. Donald L. Mallick and Peter W. Merlin, *The Smell of Kerosene: A Test Pilot's Odyssey* (Washington. DC: NASA SP-4108, 2003).

a Lockheed F-104 Starfighter, a single-engine jet with stubby wings, a T-tail, and Mach 2 cruise capability.

On November 29, 1965, Walker flew his first XB-70 chase mission in NASA 813, one of three F-104N models specially built for the Agency. The F-104N was essentially identical to the F-104G export model, but with weapons system removed, additional fuel tanks installed in the gun and ammunition bays, and an MH-97 autopilot and LN-3 navigational systems installed. During its 3-year service life, NASA 813 was flown 409 times by nine different NASA pilots, logging 627.7 flight hours. Walker made 165 of these flights, including 4 additional XB-70 chase flights in November and December 1965 and 7 more in March 1966.[13]

Decision Chain

In the spring of 1966, a series of events led to tragedy at Edwards AFB and the NASA Flight Research Center. Officials at North American Aviation, GE, NASA, and the Air Force made decisions that ultimately contributed to the loss of two aircraft and two experienced test pilots, as well as serious injury to a third test pilot.

John M. Fritz, chief test pilot for GE, set the chain of events in motion by requesting permission from a North American Aviation representative for a publicity photo session involving a formation flight of five airplanes powered by GE engines. Subsequently, XB-70 Test Director Col. Joseph Cotton agreed to provide such an opportunity on a noninterference basis following a regularly scheduled test flight.

The first opportunity to avert tragedy came when a North American official disapproved the photo session request, citing a tight testing schedule. Cotton and Fritz, however, lobbied to include it immediately following an upcoming test mission. North American officials and Cotton's immediate supervisor, Col. Albert Cate, finally approved the photo opportunity, but further approval was not sought from higher headquarters authorities, as would have been in keeping with standard procedure for such requests. Authorities might have disapproved the formation flight as an unnecessary risk.[14]

Several people were involved in planning the photo mission. Cotton arranged to include an Air Force T-38A, already scheduled as a chase aircraft for the XB-70, and Fritz requested a Navy F-4B from Point Mugu Naval Air Station, on the Southern California coast. Naval authorities approved the event as a routine training flight in support of what was assumed to be an

---

13. Merlin and Moore, *X-Plane Crashes.*

14. XB-70/F-104 Accident Investigation Report, 1966, NASA Dryden Historical Reference Collection.

The photo opportunity took place at the end of a scheduled test mission. A Learjet (at right) carried motion picture and still photographers. The pilot of the F-104D from which this picture was taken reported that the formation looked good, although the two aircraft off the bomber's left wing were not flying as close as those on the right. (U.S. Air Force)

approved Air Force mission. Fritz himself planned to fly a YF-5A, bailed to GE by the Air Force. Although his officially stated purpose for the flight was to conduct engine airstart evaluations, it was later determined that he never actually performed such tests. Fritz also tried to arrange for a B-58 to join the formation, and for supplementary Air Force photo coverage, but he was unsuccessful. GE officials contracted with charter pilot Clay Lacy for the use of a Learjet photo plane, and Cotton requested a NASA F-104 from the Flight Research Center. As NASA chief pilot, Joe Walker was within his authority to schedule the flight as a chase operation, but his superiors were unaware of the photographic mission.

An Air Force Public Affairs officer in Los Angeles learned of the photo opportunity just 2 days before the flight through a call from a commercial photographer. He referred the caller to Col. James G. Smith, Chief of Public Affairs at Edwards AFB. Smith also had been unaware of the planned formation flight but voiced no objections once he ascertained that Cate had approved the mission. The stage was now set for disaster.[15]

Midair Collision

The XB-70 test flight scheduled for June 8, 1966, had three primary objectives, the first of which was a series of subsonic airspeed calibration runs. The second

---

15. Ibid.

objective was to familiarize Maj. Carl Cross, recently assigned to the XB-70 program, with the aircraft's basic handling qualities. A third objective—added the day prior to the flight—was a supersonic speed run for the purpose of creating a sonic boom in support of SST research objectives.

North American officials held a preflight briefing on June 7 to discuss test objectives and procedures. Fritz then briefed the photo mission, describing the planned formation as a loose V-shape led by Al White and Cross in the XB-70. Walker was assigned to fly his F-104 at the inboard position, just off the bomber's right wing. The Navy F-4, piloted by Cmdr. Jerome P. Skyrud, chief of the Air Weapons Branch of the Naval Missile Center, was to be positioned off the left wing. The outboard positions, right and left respectively, were to be flown by Fritz in the F-5 and Capt. Peter C. Hoag in the T-38, with Cotton in the rear seat of the latter. The Learjet, carrying motion picture and still photographers, would remain separate from the formation, maneuvering as necessary to provide the best photo opportunities. Detailed discussion of how the formation would be flown failed to include specific separation distances. No formation commander was designated, but Cotton considered himself to be in charge of the photographic mission.[16]

The formation included (left to right) a T-38A, F-4B, XB-70A, F-104N, and YF-5A. Note the proximity of the F-104 to the bomber's right wingtip. (U.S. Air Force)

---

16. Ibid.

Al White was not present at the briefing, nor was Skyrud, whose only briefing consisted of telephone conversations with Fritz. Learjet pilot Clay Lacy was also absent but was briefed by Fritz shortly before takeoff on the morning of the flight. To make up somewhat for this deficiency, a GE test pilot who had been at the preflight briefing and was familiar with flight procedures in the Edwards AFB area accompanied Lacy as an observer. Cross had attended the briefing, as had the Learjet copilot.

XB-70 Program Director John McCollom had been the highest-ranking Air Force official at the preflight briefing. Upon hearing the details of the photographic mission, he voiced no objection, though it was within his power to cancel it.

Following takeoff on the morning of June 8, White and Cross executed the planned tests without difficulty. Although the original flight plan called for the formation to rendezvous in the vicinity of Lake Isabella, north of Bakersfield, CA, a buildup of cumulus clouds in the area necessitated moving eastward to the area near Three Sisters Dry Lake, north of the town of Barstow, CA. This required a change in altitude from 20,000 to 25,000 feet, as well as a change in direction from north-south to east-west, resulting in a shorter flightpath than originally briefed. One by one the aircraft joined the formation, flying a left-hand racetrack pattern at 300 knots indicated airspeed, with the Learjet usually to the left and about 600 feet behind the others.[17]

Ground controllers kept the formation apprised of the presence of other air traffic in the area. An Air Force F-104D, a two-place aircraft with a photographer in the back seat, received permission to join the formation for a few pictures before returning to Edwards AFB after a separate mission. The pilot of the F-104D later reported that the formation looked good, although the two aircraft on the left were not flying as close to the XB-70 as those on the right.

Because the Learjet was not equipped with ultra-high-frequency radio, as were the military aircraft, North American flight controller Frank Munds relayed communications between Lacy and the flight formation from a ground station at Edwards AFB. The photo session had been scheduled to last 30 minutes, after which Fritz asked Munds to contact the Learjet and see if the photographers were finished.

"I think we've given them as much as they expected," said Fritz.

After relaying the question, Munds replied, "The Learjet said they're still taking pictures, and they'll let us know when they get through." Cotton added that Lacy had requested an additional 15 minutes, if possible.[18]

---

17. Ibid.

18. Ibid.

Relative positions of the F-104 and the right wingtip of the XB-70 prior to impact. The strength and severe gradients of the vortex flow around the bomber's large, highly swept wing made it difficult for the pilot of the smaller aircraft to maintain safe separation during close-formation flight. (U.S. Air Force)

Each member of the formation verified having enough fuel to remain aloft a little longer. About 10 minutes later, White noticed a contrail indicating an aircraft approaching from the east. Lacy advised the formation that the photo session would be completed within about 4 minutes. Air traffic control personnel at Edwards AFB reported that the approaching aircraft was a B-58 on a speed run. It was on course to pass 5,000 feet above the formation and would pose no hazard.

Disaster struck without warning. The tail of Walker's F-104 collided with the Valkyrie's right wingtip, allowing the horizontal tail surfaces of the Starfighter to be caught by a swirling vortex in which high-pressure air from the lower surface of the bomber's wing met low-pressure air from the upper surface. The F-104 pitched upward, tearing through the XB-70's right wingtip. The spinning airflow of the vortex then rolled Walker's plane to the left, over the top of the bomber's fuselage, where it sheared off both of the Valkyrie's vertical tails. By this point the F-104 was in several pieces, the largest engulfed in a ball of fire fed by ruptured fuel tanks. Walker was killed instantly.

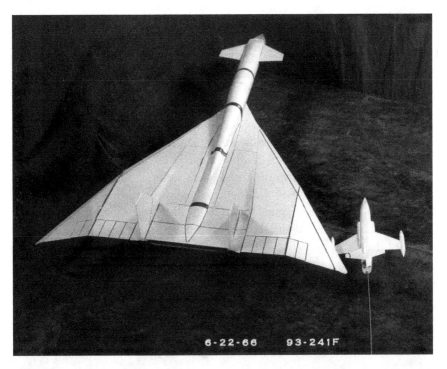

First contact between the F-104 and XB-70 is depicted using scale models. The upper surface of the Starfighter's left horizontal stabilizer struck the lower surface of the Valkyrie's right wing-tip. The F-104 pitched upward, tearing through the bomber's wing before rolling over the top. (U.S. Air Force)

Fritz, flying off Walker's right wing, saw nothing amiss prior to the collision. Immediately after impact, the XB-70 continued in straight and level flight for 16 seconds as though nothing had happened. Voices on the radio frantically cried, "Midair! Midair! Midair!"[19]

Cotton informed the XB-70 crew, "The verticals came off, left and right. We're staying with you. No sweat. Now you're looking good."

Just then, the Valkyrie rolled ponderously to the right and entered an inverted spiral. As it shed parts, fuel poured from its broken right wing. Cotton and others began calling for the stricken crew to eject. "Bail out! Bail out! Bail out!"

Al White initiated his ejection sequence by squeezing the right-side hand-grip to encapsulate his seat, but his right elbow got caught as the clamshell doors closed. By design, the door closure power impulse was less than 15 pound-seconds so that serious injury would not result in the event of limb entrapment,

19. Ibid.

but this force was still equivalent to the force of a 45-pound object falling from 1.5 feet. Additionally, the pilot's trapped arm served as a compression link between the upper clamshell door and the raised handgrip, effectively preventing completion of the encapsulation sequence. While struggling to free his arm, White opted to eject with the capsule doors open, which was a survivable procedure at the existing speed and altitude conditions. He successfully executed hatch jettison and ejection by squeezing the left-side handgrip. Seconds later he managed to free his elbow, wrenching his arm in the process.[20]

The capsule's recovery parachute deployed, but an impact attenuator bag on the underside of the capsule failed to inflate because the clamshell doors were still open. White closed the doors manually but, disoriented and in pain, he forgot to activate the manual inflation handle before the capsule struck the ground with an acceleration of 45 g's. Fortunately, the seat structure absorbed some of the force, but White suffered severe bruising. Nevertheless, he managed to egress the capsule unassisted.

Cross, possibly incapacitated by extreme g-forces, was less fortunate. His seat retraction mechanism apparently failed, making automatic encapsulation impossible. He remained trapped in the stricken jet and died when it struck the ground.[21]

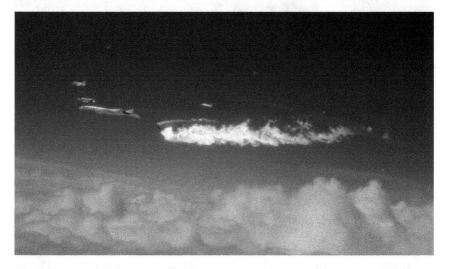

The spinning airflow of the vortex rolled Walker's plane to the left, over the top of the bomber's fuselage, where it sheared off both of the Valkyrie's vertical tails. By this point, the F-104 was in several pieces, the largest engulfed in a ball of fire from its ruptured fuel tanks. Walker was killed instantly. (U.S. Air Force)

20. Ibid.
21. Ibid.

Black columns of smoke darkened the sky over the debris-strewn desert landscape. Two experienced test pilots were dead and another badly injured. The cost of the XB-70 was estimated at $217.5 million, including its six engines and other Government-furnished equipment. NASA reckoned the loss of the F-104N to include its $2 million acquisition cost, plus the cost of special instrumentation. An Accident Investigation Board was formed immediately to determine probable causes and make recommendations.[22]

Aftermath

Based on photographic and other evidence, investigators concluded that the swirling wake vortex from the bomber's wingtip became a contributory factor in the accident only after the F-104's tail came so close to the XB-70 that a collision was imminent.

Walker had a reputation as a fine test pilot and a levelheaded professional, dedicated and safety conscious. He had nearly 5,000 hours of flight experience and had previously flown chase for the XB-70 nine times, eight of those in an F-104. It was hard to imagine that lack of formation proficiency or some lapse of judgment could have led to the disaster. Investigators decided it was most likely that Walker had somehow become distracted and inadvertently moved his control stick, causing the F-104 to move imperceptibly toward the XB-70.[23]

Shortly before the collision, after air traffic controllers reported the approaching B-58, several pilots in the formation responded that they could see the B-58's contrail. Walker never made such a call. He may have been attempting to spot the B-58 at the time his aircraft collided with the XB-70.

The Accident Investigation Board ultimately concluded that Walker's position relative to the XB-70 left him with no good visual reference points for judging his distance from the bomber. Therefore, a gradual movement in any direction would not have been noticeable to him. The board determined that an "inadvertent movement" of the F-104 had placed it in a position such that contact was inevitable.[24]

The length of the precision formation mission may have also been a factor. Cloudy weather had extended the flight time and forced the formation to move to a different area than originally planned. Additionally, Walker had been flying close to the bomber's wing in a position that made it difficult for him to judge his distance from it. Likewise, other air traffic in the area created distractions. The effort to maintain formation created a heavy cockpit workload for Walker,

---

22. Merlin and Moore, *X-Plane Crashes*.

23. Mallick and Merlin, *The Smell of Kerosene*.

24. XB-70/F-104 Accident Investigation Report, p. 5.

further complicated by his proximity to the Valkyrie's wingtip. At one point he told Cotton, "I need another set of hands here, Joe."[25]

A number of people suffered the consequences for their role in organizing, planning, and approving the flight. A collateral investigation board ruled that Col. James Smith, as head of the Information Office at Edwards AFB, should have advised the responsible parties of proper procedures for approving such a mission through higher headquarters authorities. Investigators found that "the Air Force officer who actively assisted GE [in arranging the photo mission] exercised poor judgment and his superior, who approved the flight with full knowledge of these arrangements, did not properly discharge his duties."[26]

In an August 1966 memorandum, Air Force Secretary Harold Brown informed the Secretary of Defense that "The photographic mission would not have occurred if Col. Cotton had refused the GE request or at least not caused North American to reconsider its reluctance. It would not have occurred if Col. Cate had taken a more limited view of his own approval authority. It would not have occurred if Col. Smith had advised of the need for higher approval. It would not have occurred if Mr. McCollom had exercised the power he personally possessed to stop the flight."[27]

Brown further stated, "From all the evidence, these individuals acted in ignorance of prescribed procedures, rather than with intent to violate them."[28] He noted that the commander of Air Force Systems Command, with the concurrence of the Air Force Chief of Staff, had directed a number of disciplinary actions against the responsible parties. Cate was relieved as deputy for Systems Test and reassigned. Cotton and Smith received written reprimands, as did McCollom.

The Air Force also made numerous administrative changes to improve operational procedures, starting with the correction of supervisory and procedural weaknesses within the responsible test organization. It was clearly essential that test directors have wide latitude to make decisions, but not at the expense of positive command and control of their operations. The board also recommended that higher headquarters authorities ensure policy compliance; Brown promised to update older regulations, simplify provisions that were unnecessarily complex, and clarify points that defied uniform interpretation.

Brown's conclusions indicated, "None of the participants in the events leading to the accident emerges in this analysis as exercising good judgment. But

25. Ibid., p. 38.

26. Ibid., p. 2.

27. Ibid., p. 10.

28. Ibid., p. 10.

the responsibility for observance of Government regulations and for assuring proper use of Government resources rested with the Air Force personnel."[29]

As with most accidents, there was no single cause for this mishap. It would be shortsighted and naïve to blame a test pilot with Joe Walker's experience and credentials for inadvertent stick input that allowed the F-104 to drift too close to the XB-70. In fact, he was merely the last link in a chain of events that established the latent conditions necessary for such a mishap to occur.

Had Lacy not requested an additional 15 minutes of formation time, the aircraft would have separated before the B-58 passed overhead, thus eliminating the potential for distraction. Had mission planners not chosen are area prone to cumulus cloud buildup for the photo opportunity, the formation would not have had to move from the Lake Isabella area to the Three Sisters area, which added pilot workload and lengthened the mission. Had McCollom voiced an objection to the photo formation during the preflight briefing, that portion of the mission would have been scrubbed. Had any of the Air Force officers involved routed the original request through proper channels, the photo opportunity would almost certainly have been disapproved. Had Cotton simply accepted the objection by North American officials, or had he turned down Fritz's request in the first place, tragedy would have been avoided.[30]

As in any system, the action or change of one part of the system can have unintended and unforeseen consequences in another. In the case of the XB-70, arranging a photo opportunity involving so many different aircraft, in such close flying proximity, arguably without an adequate prebriefing and without due oversight from superiors, was one such example. In effect, they were treating the XB-70 as an operational aircraft, rather than the experimental test aircraft that it actually was. Organizational safety researcher James Reason likens the layers of supervision and management in an organization to slices of cheese—specifically, Swiss cheese. The holes in each slice of the cheese represent areas of safety vulnerability in the context of operations. When the holes are small and out of alignment with one another, safe operations ensue. But when the holes in the various layers line up, the "accident arrow" is allowed to pass through, resulting in an accident. In this case, the accident was a tragedy. Such are the lessons to be learned from the XB-70 mishap.[31]

---

29. Ibid.

30. Ibid.

31. Reason, *Managing the Risks of Organizational Accidents.*

The B-1 was originally designed to operate at near-sonic speeds at treetop heights and at supersonic speeds at high altitude. The engines were located close to the aircraft's center of gravity in order to provide the greatest stability in the turbulent conditions expected during low-altitude penetration attacks. (U.S. Air Force)

CHAPTER 8

Mission Management and Cockpit Resource Management in the B-1A Mishap

Contrary to popular myth, flight research does not involve a lone hero strapping himself into the cockpit and taking off on his own to push the edge of the envelope. Numerous people are involved in carefully planning each test point to be performed on a mission, scheduling test activities in a logical sequence and directing and managing the conduct of the mission. Testing a highly complex aircraft requires the coordinated efforts of mission planners, test pilots, flight-test engineers, and chase crews in the air, as well as test directors, test conductors, and engineers on the ground. Mission management and crew resource management (CRM) are critical to the safe execution of a test flight.

Advanced Strategic Bomber

In 1961 the Department of Defense, recognizing the need to modernize the Air Force's strategic bomber force, established a requirement for a low-altitude penetration bomber to replace the Boeing B-52 by 1980. Subsequently, in June 1970, the Advanced Manned Strategic Aircraft study culminated in the selection of Rockwell International for the construction of a new bomber called the B-1. Plans to order five prototypes were modified in January 1971 in a cost-cutting exercise that eliminated two of the flying prototypes, as well as one static test specimen. Later, one of the prototypes was restored, allowing work to proceed on four, while plans were made to order 240 operational B-1 bombers.

The airplane's configuration featured a blended wing-body structure with variable geometry outer wing panels. For low-speed flight maneuvers such as takeoff and landing, the wings were swept forward to maximize lift. For high-speed cruise, the wings were swept aft to reduce drag. Principal materials used in construction of the B-1 included aluminum alloys and titanium—the latter used primarily in the wing carry-through structure, engine nacelles, and aft

The B-1's configuration featured variable geometry outer wing panels. For low-speed flight maneuvers such as takeoff and landing, the wings were swept forward to maximize lift. (U.S. Air Force)

fuselage. The balance of the airframe was made up of steel and a number of nonmetallic materials.[1]

Depending on mission requirements, the pilot could vary the wing-sweep angle from 15 degrees (maximum forward sweep) to 67.5 degrees aft. Changing the position of the wings and flaps altered the aircraft's static margin, which is the distance between the center of gravity and the center of lift, therefore affecting its stability. In order to maintain center of gravity within flyable limits, the B-1 was equipped with a system for automatically or manually transferring fuel between different tanks. In either mode it was critical for the flightcrew to closely monitor the center of gravity during wing-sweep maneuvers.[2]

The B-1 was originally designed to operate at near-sonic speeds at treetop heights and at supersonic speeds at high altitude. It was equipped with four GE YF101-GE-100 afterburning turbofan engines mounted in pairs beneath the fixed inboard section of the wing. The engines were located close to the center of gravity in order to provide the greatest stability in turbulent conditions expected during low-altitude penetration attacks.[3]

1. Don Logan, *Rockwell B-1B: SAC's Last Bomber* (Atglen, PA: Schiffer Military History, 1995).
2. Investigation of USAF aircraft accident, August 29, 1984, B-1A 74-0159 (Air Force Flight Test Center, 1984).
3. Logan, *Rockwell B-1B*.

For use in the event of a serious emergency, the B-1 was equipped with an escape capsule. The entire crew cabin was designed to separate from the fuselage using two rocket motors. Three parachutes lowered the capsule to the ground while attenuator bags inflated to reduce impact forces. In the event of water impact, the attenuator bags served as flotation devices. However, this configuration was used only on the first three airframes. A fourth prototype was equipped with standard ejection seats, as were planned for the production model.[4]

Construction of the first three B-1 prototypes began in late 1972 at Rockwell's facility in Palmdale, CA. The first, known as Air Vehicle 1, or AV-1, was completed in October 1974. The maiden flight from Palmdale to nearby Edwards AFB took place on December 23, crewed by Rockwell test pilot Charles C. Bock, Jr., Col. Emil "Ted" Sturmthal, and flight-test engineer Richard Abrams. This marked the beginning of developmental testing by the B-1 CTF, composed of personnel from Rockwell, Air Force Systems Command, and Strategic Air Command. Test flights were scheduled a few weeks apart in order to allow engineers to assimilate extensive amounts of data and integrate the results into plans for subsequent flights.

The second prototype, AV-2, served as a structural test aircraft. Following static loads testing, the airframe was inspected, rebuilt, and readied for flight testing. Tommie D. "Doug" Benefield piloted its first flight on June 14, 1976. A few years later, he accelerated AV-2 to Mach 2.22, the highest speed attained by any B-1 during the test program. Two more prototypes joined the fleet, but AV-2 was to have the most extensive and varied career.[5]

In June 1977, President Jimmy Carter ordered B-1 production canceled, though research and development tests using the four prototypes were allowed to continue at Edwards AFB. The flight program ended on April 29, 1981, with a final flight of AV-4. By this time, the test team had flown a combined total of 1,895 test hours.

In January 1982, President Ronald Reagan decided to reinstate B-1 bomber production. Under a $1.3 billion contract, Rockwell agreed to produce 100 aircraft but with significant differences from specifications for the original prototypes. The configuration of the B-1B included ejection seats (as had been planned for the production B-1A), redesigned engine inlets and wing fairings, and improved avionics.[6]

AV-2 was subsequently assigned to support the B-1B test program and underwent modifications that included the installation of new weapon bay

---

4. Steve Pace, *Boeing North American B-1 Lancer* (North Branch, MN: Specialty Press, 1998).

5. Logan, *Rockwell B-1B*.

6. Ibid.

doors, changes to some internal structures, and the addition of a B-1B-type FCS. The modified aircraft made its first flight on March 23, 1983.

During the B-1B test program, AV-2 was used to conduct stability, control, flutter, and weapon separation tests. The aircraft completed initial separation tests with the Short Range Attack Missile, Mk.82 conventional high explosive bomb, and B61 and B83 nuclear weapon shapes. Additional plans called for modifying AV-2 to carry the AGM-86B air-launched cruise missile, scheduled for testing in mid-1985. During its service life, AV-2 was flown 60 times during the initial B-1 flight-test program and 66 times during the B-1B test program, logging a total of 543 flight hours. More flights were planned, but tragedy intervened.[7]

Mixed Experience

The 127th flight of AV-2 was scheduled for August 1984 and designated Flight 2-127. Mission objectives included taxi tests, takeoff performance evaluation, airspeed calibration, weapon separation tests with Mk.82 500-pound

inert practice bombs, and minimum-control-speed tests to evaluate the bomber's low-speed handling characteristics in various configurations. On the day prior to the mission, Rockwell test engineer Warren Kerzon held a Flight Readiness Review (FRR) briefing to identify the aircrew and test personnel involved; discuss aircraft status; and outline requirements for instrumentation, tanker support, chase aircraft, and range scheduling. He also reviewed test cards that described all planned maneuvers and data points. Present at the meeting were the B-1 Combined Test Force (CTF) director and Rockwell's flight-test engineering manager, aircraft manager, and acting director of flight-test operations. Test and chase aircrew members were also present and participated in a crew briefing following the FRR.

Maj. Richard V. Reynolds was assigned as pilot in command of Flight 2-127. Prior to the mishap flight, his overall experience in the B-1 consisted of 13.9 flight hours during three training sorties, plus a final checkout. (U.S. Air Force)

---

7. Ibid.

Maj. Richard V. Reynolds was assigned as pilot in command with Rockwell chief test pilot Doug Benefield as copilot and Capt. Otto J. Waniczek as flight-test engineer. Each crewmember was qualified for the mission, but the difference in experience between the pilot in command and the copilot played a part in the events to follow.

Reynolds had graduated from the Air Force Academy in 1971 with a bachelor of science degree in aeronautical engineering. He attended undergraduate pilot training at Vance AFB, OK, in 1971 and 1972, remaining at Vance as a T-38 flight instructor and check pilot until November 1975. Reynolds then transferred to Fairchild AFB, WA, as a B-52G aircraft commander, instructor pilot, and flight examiner. In 1978 he was a distinguished graduate of the Squadron Officer School at Maxwell AFB, AL. He then attended the U.S. Air Force Test Pilot School at Edwards AFB, completing the course as a distinguished graduate in June 1980. Following graduation, he was assigned as an experimental test pilot and project pilot with the B-1 CTF. In June 1981, he took on the additional duty of chief of flight safety at Edwards. In August 1982, he went to the Air Command and Staff College at Maxwell AFB, AL, emerging as a distinguished graduate in July 1983. He returned to Edwards AFB as project pilot and operations officer with the B-1B CTF. By September 1984, he had accrued more than 3,000 hours of flight time in numerous aircraft types, including jet fighters and bombers, transports, trainers, general-aviation types, helicopters, and sailplanes.[8]

As pilot in command of Flight 2-127, Reynolds was responsible for the conduct of the entire flight up to, but not including, a segment involving landing re-currency qualification, during which Benefield would serve as instructor pilot. To maintain currency, each pilot had to make one landing every 60 days, and Reynolds was 9 days overdue. His overall experience in the B-1 consisted of 13.9 flight hours during three training sorties, plus a final checkout. With the exception of landing currency, Reynolds was considered fully qualified for the mission.[9]

Benefield joined the Air Force in 1949, immediately after graduating from Texas A&M University with a bachelor of science degree in aeronautical engineering. He earned his pilot wings in August 1950 and subsequently flew combat missions over Korea. Following the Korean War, he served as a transport pilot at Sewart AFB, near Smyrna, TN. Benefield graduated from the Test Pilot School at Edwards AFB in 1955 and the Test Pilot School Aerospace Research Pilot

---

8. USAF Leadership Biography, Lt. Gen. Richard V. Reynolds, *http://www.af.mil/information/bios/ bio.asp?bioID=6888*, accessed November 19, 2009.

9. Investigation of USAF aircraft accident, B-1A 74-0159, August 29, 1984.

School (Space Course) in 1962. He was a member of the Group 3 Air Force astronaut designees announced in October 1962, placing him in line to fly the planned X-20 Dyna-Soar spaceplane. Following the cancellation of the Dyna-Soar project, he remained at Edwards AFB as a test pilot until 1966. There he evaluated the stall characteristics of the C-133 cargo plane. According to fellow test pilot Fitzhugh "Fitz" Fulton, "His work on the C-133 saved the lives of many people on later crews."[10]

With his extensive experience in bomber and transport operations, Benefield was assigned by the Air Force to the FAA as a test pilot for the supersonic transport development program. He also assisted with the certification of the

British/French Concorde American SST. Later, during the Vietnam War, he flew 176 combat missions over Southeast Asia in the F-4. After retiring from the Air Force in 1973, Benefield joined Rockwell International as a test pilot for the B-1. He was appointed the company's chief test pilot in 1983. Benefield logged over 11,000 hours in more than 50 types of aircraft. He was a fellow of the Society of Experimental Test Pilots and served as its president in 1983. He was awarded the American Institute of Aeronautics and Astronautics Octave Chanute Flight Award in 1977.[11]

Benefield's responsibilities for Flight 2-127 included serving as copilot during the various phases of the test mission. For landing, he would act as instructor pilot for Reynolds's landing re-currency qualification. He was fully qualified to execute these duties, according to Maj. Richard L. Bates, chief of Standards/Evaluation for the

Tommie D. "Doug" Benefield was assigned as copilot for most phases of the test mission. For landing, he was to act as instructor pilot for Reynolds's landing re-currency qualification. At the time of the mishap, Benefield was considered the most experienced and knowledgeable B-1 pilot in the world. (U.S. Air Force)

---

10. *Aerospace Walk of Honor* (Lancaster, CA: Davis Communications, 2009).
11. Ibid.

6510th Test Wing, who noted that Benefield "was probably the most experienced and knowledgeable B-1 pilot in the world."[12]

Waniczek had been stationed at Edwards AFB for more than 7 years. In 1977, he joined the original B-1 CTF as a flying qualities engineer. Two years later, he transferred to the 6513th Test Squadron to conduct flying qualities evaluations of various fighter jets. In January 1983, he graduated from the Air Force Test Pilot School as a flight-test engineer and was then assigned to the B-1B test program as an operations engineer. By 1984, he had logged more than 340 flight hours as test engineer in a wide variety of aircraft. Of all flight-test engineers qualified in the B-1, he was considered the most experienced.

Waniczek's duties in the CTF included responsibility for scheduling, planning, and conducting missions in AV-2. He was responsible for operating onboard test instrumentation during Flight 2-127, along with various other engineering tasks. With over 60 flight hours in the B-1, he was considered fully qualified to carry out his duties.[13]

Ground Monitoring

A variety of other personnel supported the test flight from the Ridley Mission Control Center at Edwards AFB. During the flight, engineering specialists analyzed strip chart recorders that displayed 80 channels of analog data in real time, including the aircraft's center of gravity. Selected aircraft parameters, including center of gravity and wing-sweep position, were displayed on four television monitors located within the room.

Unfortunately, there were no indicators to highlight unsafe aircraft parameters. The flightcrew had onboard caution lights and alarms, but such information was not available to mission controllers on the ground. The original B-1 flight-test data-instrumentation system had automatically provided flight safety data to engineers on the ground, but this equipment was retired prior to the B-1B program. The system available for Flight 2-127 required direct, individual monitoring of safety parameters by mission controllers and direct radio communication to confirm or resolve discrepancies. No individual was specifically tasked with overall responsibility of ensuring that center of gravity and wing-sweep position were within acceptable tolerances during the test mission.[14]

Robert Broughton, Rockwell's manager of flight-test engineering, served as test director for Flight 2-127. He was responsible, along with the pilot in

---

12. USAF investigation, B-1A 74-0159.

13. Ibid.

14. Ibid.

command, for all real-time decisions throughout the flight. These included modifying the flight plan or aborting the mission if conditions warranted. He had access to all control-room personnel and a television monitor displaying aircraft parameters. Broughton, a former test pilot, had been involved with the B-1 program as an Air Force officer in the B-1 Systems Program Office. After retiring from military service in 1976, he was hired by Rockwell to participate in B-1 test planning activities.

Maj. Stephen Henry, a qualified B-1 crewmember, was the designated test conductor for the flight. He served as the single point of direct communication with the aircrew. He was also responsible for reviewing the quality of data at each test point and communicating to the aircrew clearance to proceed to the next set of test conditions. After entering the Air Force in 1972, Henry served as a navigator on B-52 bombers and was later involved in testing avionics, navigation systems, and radar warning systems for the A-10. Following graduation from the Test Pilot School in 1982, he was assigned to the B-1 test program.[15]

Between Test Points

At 5 a.m. on August 29, Reynolds, Benefield, and Waniczek received a weather briefing and aircraft status update. The original flight plan called for airspeed calibration tests in the vicinity of the airfield immediately after takeoff, followed by static and dynamic minimum-control speed tests in the Cords Road area northeast of Edwards AFB. Then, following in-flight refueling, the bomber would proceed to the Edwards Precision Impact Range Area for the release of five concrete-filled Mk.82 high-drag bombs. Reynolds would then pilot several touch-and-go landings to establish re-currency, as well as a landing touchdown load test and a maximum-braking effectiveness test.

The mission actually began before takeoff, with ground-load-survey turning tests performed during taxi en route to the runway. As a result of excessive wear, several tires had to be replaced, delaying takeoff until about 9:30 a.m. Because of this delay, the test director rescheduled weapon separation tests for immediately after takeoff, followed by airspeed calibration and the first set of minimum-control speed tests. The final sequence of minimum-control speed maneuvers was scheduled following aerial refueling. Landing events remained unchanged.[16]

The mission began without incident. Only two airspeed calibration runs were completed because thermal turbulence in the flyby pattern resulted in questionable data. Reynolds and his crew then proceeded to climb to an altitude of 6,000 feet mean sea level (MSL) (roughly 4,000 feet above the ground),

---

15. Ibid.

16. Ibid.

Depending on mission requirements, the pilot could vary the wing-sweep angle from 15 degrees (maximum forward sweep) to 67.5 degrees aft. It was critical for the flightcrew to closely monitor the center of gravity during wing-sweep maneuvers in order to maintain the center of gravity within flyable limits. (U.S. Air Force)

and flew north to the Cords Road Test area to prepare for minimum-control speed tests. Cords Road is an unpaved track running east from California City and passing north of Harper Dry Lake, about 35 miles northeast of Edwards AFB. The sparsely inhabited area is ideal for low-altitude flying since there is little risk to civilians or private property.

While flying an easterly course, Reynolds and Benefield began configuring the aircraft for the first set of data points. For the static test points, the wings were swept aft 55 degrees and the center of gravity set at 45 percent mean aerodynamic chord (MAC), the average distance from the leading edge of the wing to the trailing edge. The airplane was in an aerodynamically clean configuration: flaps, slats, and landing gear retracted.[17]

Reynolds had not previously flown any test points in the B-1 with aft wing sweep, so he had mild concerns about rolling the aircraft during low-speed maneuvers. On the first attempt he left the speed brakes open, making it necessary to repeat the first test point. He ultimately completed both static

---

17. Ibid.

points without incident, although there was considerable discussion between the aircrew and mission control regarding the quality of the data and whether a lower aircraft gross weight might have produced better results.

The B-1 was equipped with a master caution panel with numerous warning lights for alerting the crew to malfunctions and potentially hazardous conditions. Warning lights occasionally illuminated throughout the flight. However, because the situations the lights signified most often were not serious, Benefield simply reset the master caution indicator each time. The crew gradually became anesthetized to the alarms in what Reynolds later described as "warning fatigue."[18]

In order to reconfigure the airplane for dynamic minimum-control speed tests, Reynolds began moving the wings forward, and he asked Waniczek for the target airspeed (minimum controllable airspeed plus 9 knots). At the time, the bomber was cruising at 300 knots calibrated airspeed. Benefield lowered the landing gear and extended flaps and slats. For reasons that were never identified, he failed to manually change the center-of-gravity setting to 21 percent MAC, as required for the test point. Meanwhile, Reynolds swept the wings to the full forward position in one continuous motion, even though he had been advised during his checkout to sweep the wings in stages.

As the airplane slowed to 200 knots, the crew heard a warning tone and saw that the master caution light and a flap caution light had both illuminated, indicating that the B-1 had two engines set below normal cruise power with the airplane flying below 10,000 feet MSL. Reynolds reached up and extinguished the master caution light, noting that the flaps appeared to be properly trimmed. There was no apparent indication of a problem with the center of gravity, although the flap caution warning light did in fact serve as a center-of-gravity limit caution under certain conditions.[19]

During a typical B-1 test flight, the aircrew coordinated and fine-tuned the center-of-gravity settings with engineers on the ground that had access to very precise telemetered data on strip-chart readouts. On this occasion, however, there was no such coordination. During the gap between static and dynamic tests, control-room personnel turned away from their strip charts and began discussing the quality of the data received during the previous test points and how it might be possible to get better data by repeating the tests at a lower gross weight, after burning off about 10,000 pounds of fuel. During

---

18. Interview with Lt. Gen. Richard V. Reynolds, USAF, ret., at NASA DFRC, Edwards, CA, July 15, 2008.
19. USAF investigation, B-1A 74-0159.

this discussion, no one noticed that the airplane's center of gravity was moving dangerously aft because fuel had not been transferred forward.[20]

A design feature that had previously been noted as a potential risk also came into play. The B-1 was equipped with a full-stabilizer-authority variable-gain stability control augmentation system (SCAS) that scheduled gains as a function of Mach number and altitude. With the pitch SCAS command-path gain set at maximum, only very light stick forces were required to fly the aircraft. Even though the bomber was decelerating, its configuration was being changed, and, due to the changing center of gravity, it was becoming aerodynamically unstable. This design characteristic masked the degrading flying qualities until the bomber became uncontrollable.

When the wings reached their forward limit, the center of gravity briefly shifted to 42 percent MAC and then stabilized at 46 percent MAC. The AOA was 8.5 degrees. With airspeed decreasing through 146 knots and the center of gravity 25 percent beyond its aft limit, disaster was imminent.[21]

Reynolds successfully initiated the crew escape sequence just 1,500 feet above the ground. As the main parachutes opened, however, a repositioning mechanism failed. Instead of landing on inflated impact-attenuation bags, the capsule struck the ground nose-down with a force of 40 g's. (U.S. Air Force)

In response to a request from the test conductor to repeat the earlier data points, Reynolds agreed to comply but did not think it would make a difference. Then, as Reynolds began to request a new target weight, he felt the bomber's nose begin to rise alarmingly. He and Benefield pushed forward on their control sticks, set throttles to full, and ignited the afterburners, but the nose continued to pitch upward to 70 degrees. Wallowing first left, then right, the aircraft shuddered and groaned, while the cockpit filled with smoke.[22]

---

20. Reynolds interview, 2008; USAF investigation, B-1A 74-0159.

21. USAF investigation, B-1A 74-0159.

22. Reynolds interview.

Rockwell test pilot Mervin Evenson, flying chase in an F-111 with Capt. Stephen Fraley, called, "Hey, Doug, have you got it?" Benefield responded, "No. We are trying to come out of it now, maybe."

As the stricken bomber plummeted toward the ground, Benefield recognized that recovery from a low-altitude stall was impossible. "We have to punch," he told Evenson.[23]

Just 1,500 feet above the ground, Reynolds initiated the sequence to separate the crew escape capsule. Twin rocket motors blasted the capsule free, and a drogue chute deployed to stabilize its trajectory. However, as the main parachutes opened, a repositioning mechanism failed to operate, preventing the capsule from attaining a horizontal attitude. Instead of landing on the inflated impact-attenuation bags, the capsule struck the ground nose-down with a force of 40 g's and bounced 12 feet into the air before finally landing in an upright position. In the high-g impact, Benefield's seat tore loose and slid forward along its mounting rails. He suffered fatal head injuries as a result. Waniczek suffered a partial collapse of both lungs and minor lacerations. During the flight, he had been strapped in with a lap belt but had removed

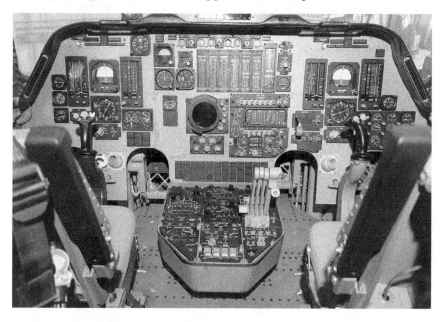

The control panel of the B-1A was equipped with a master caution panel with numerous warning lights that would alert the crew to malfunctions and potentially hazardous conditions. Since these lights occasionally illuminated due to situations that frequently were not serious, the crew gradually became anesthetized to the alarms in what is described as "warning fatigue." (U.S. Air Force)

23. USAF investigation, B-1A 74-0159.

his helmet and shoulder harness in order to facilitate movement within the cramped engineer's station. Reynolds survived the accident with back injuries and had to overcome temporary paralysis. The B-1 was completely destroyed on impact. Fortunately, due to the remote location, there were no injuries to bystanders or damage to private property. The bomber's value was estimated to be approximately $325 million.[24]

Accident Investigation

The Accident Investigation Board determined that the failure to manually transfer fuel allowed an out-of-trim condition to develop during the course of sweeping the wings forward. Once the center-of-gravity limit was exceeded, the aircraft stalled without sufficient altitude available for recovery.[25]

Lessons learned from the B-1 accident can be applied to current flight-test programs. The first relates to the importance of crew composition. During Flight 2-127, the least experienced B-1 pilot was assigned as aircraft commander with the most experienced crewmember assigned as copilot. This inequity affected efficient CRM because Reynolds tended to defer to Benefield and assume he was fulfilling his responsibilities. Such deference can lead to what Reynolds has described as "silent incapacitation." Reynolds assumed that Benefield had reset the center-of-gravity controls before the wings were swept forward. In a deviation from flight-manual procedures, Benefield did not reset the controls and neither pilot verified that the center of gravity was within limits before sweeping the wings. Additionally, crew assignments for the mission deviated from a regulation specifying that any crewmember designated as instructor pilot for any portion of the flight is considered to be in command of the aircraft at all times, regardless of which seat is occupied by the instructor pilot. Although Benefield was in the copilot's seat, he should have been designated pilot in command since he was to perform instructor duties during the landing phase of the mission. Such action would likely have thwarted the excessive deference seen in the mishap sequence.[26]

A second lesson concerned mission management and planning. "We set a trap for ourselves," Reynolds observed, "in going from a clean, aft-sweep configuration to a forward-sweep, dirty configuration with a dramatically different center of gravity." Had the test sequence called for an intermediate configuration between the two test points, there would have been a greater

---

24. Reynolds interview; USAF investigation, B-1A 74-0159.

25. USAF investigation, B-1A 74-0159.

26. Reynolds interview; USAF investigation, B-1A 74-0159.

Although Benefield was in the copilot's seat, he should have been designated pilot in command since he was to perform instructor duties during the mission's landing phase. Regulations specified that any crewmember designated as instructor pilot for any portion of the flight is considered to be in command of the aircraft at all times, regardless of which seat is occupied by the instructor pilot. (U.S. Air Force)

opportunity to properly manage fuel transfer in order to move the center of gravity forward.[27]

Third, warning fatigue led the crew to habitually ignore caution lights if they seemed inconsequential. However, even warnings that could indicate the status of center-of-gravity parameters were ignored. This was an example of what is known in the field of human factors engineering as habit pattern transfer. In these instances, a suboptimal pattern of action (or reaction) becomes established by the repeated activation of warning alarms that are of little consequence. Later, when a warning of more serious consequence is manifest, the same habit pattern of simply turning off the alarm predictably ensues.

A human factors approach to aircraft engineering and design can help reduce such hazards. For example, a cockpit caution annunciator panel configuration in which information concerning aircraft status is displayed in a tiered system,

27. Reynolds interview.

such as information (no alarm), caution (mild alarm), and warning (strong alarm), would significantly mitigate the problem of habit pattern transfer.

In retrospect, according to Reynolds, this mishap highlighted another area of concern: communication. The crew did not discuss the warnings and failed to properly interpret what the aircraft was trying to communicate to them through the master caution panel.[28]

Additionally, interaction between the cockpit and control room broke down during a critical phase of the flight. Because the wing-sweep maneuver took place between two test points, mission controllers mistakenly considered this phase of the flight to be without significant risk; a brief meeting between the test director, test conductor, and various engineers left no one to monitor the center-of-gravity values on telemetry displays, and another opportunity to warn the crew was therefore lost. Following the mishap, control-room procedures were changed to prevent a future recurrence of such incidents.

Finally, attempts by the flightcrew to regain control of the aircraft resulted in a delayed decision to eject. Typically, if a stricken aircraft is below 10,000 feet, ejection should be immediate. Loss of control in the B-1 began at an altitude of around 4,000 feet. Reynolds and Benefield waited 29 seconds before initiating the escape sequence at 1,500 feet above the ground. An earlier decision to eject would have likely reduced overall acceleration forces experienced by the crew, with a potential for less injury and greater survivability.

As pilot in command, Reynolds took full responsibility for the accident despite the fact that there were clearly a number of factors involved that were beyond his control. "We skipped a step that we always took, and that gross mistake had dire consequences."[29]

The lessons learned regarding cockpit discipline, adherence to protocol, and attention to detail are obvious.

---

28. Ibid.
29. Ibid.

Prior to construction of the International Space Station, the Russian Mir station was the only existing platform capable of providing long-duration space mission experience for U.S. and Russian crews. Between 1995 and 1998, seven NASA astronauts spent 989 days aboard Mir conducting a wide variety of scientific experiments. (NASA)

CHAPTER 9

Collision in Space: Human Factors in the Mir-Progress Mishap

Whether wintering at an Antarctic research station or serving a 6-month tour of duty on board an orbiting space station, humans exposed to prolonged isolation in a confined environment face daunting physical and psychosocial challenges. In such situations, individual performance and group dynamics can have life-or-death consequences. International exploration programs often face the additional challenges of cross-cultural differences and language barriers.

This case study examines conditions on board the Mir space station during a joint United States–Russian mission in 1997 that led to a collision with an uncrewed Progress-M spacecraft during a manually controlled docking attempt. Lessons learned from this event ideally will benefit future astronaut crews and mission planners by reducing risk to human life and critical space hardware.

The June 1997 collision between the Progress 234 supply ship and the Russian Mir station resulted from a host of systems and human factors issues that had nearly led to an identical mishap months earlier. Only quick action by the crew stabilized the very dangerous situation and obviated the need to abandon the station.

The perfect storm of human factors issues that led to the mishap serves to illustrate Reason's Swiss cheese model in which safety holes at multiple levels line up to result in a mishap.[1] Organizational influences in the accident included resource management, organizational climate, and operational processes. Contributory elements at the supervisory level included supervisor error (violation of mission rules), planned activities that were inappropriate under the circumstances, and failure to correct known problems.

System-level factors that increased the probability of a collision included pressure on the crew from mission control officials to perform a difficult task, a hectic pace of operations, previous close calls, and a variety of onboard system

---

1. Reason, *Managing the Risks of Organizational Accidents.*

failures that had left the crew in a stressed and fatigued state. Additionally, system failures in the docking apparatus itself made the attempted maneuver more difficult than usual. Finally, displays and controls had been inadequately designed for the task environment.

The general deterioration of economic conditions in Russia and the Ukraine (site of the Baikonur Cosmodrome and Russian cosmonaut training center) also appears to have played a latent organizational role. Such factors included a mission payment/incentive structure that required the crew to complete certain tasks, as well as problems with obtaining automatic docking devices from the Ukrainians for a reasonable price.

In addition, physical and psychological factors played a role: chronic fatigue from lack of sleep for an extended period prior to the mishap; use of a poor display-control interface; lack of recent docking practice; tension between the mission commander on board Mir and ground-based mission controllers, as well as psychological and financial pressures resulting from a previous docking failure with Progress 233; and other incidents.

Shuttle-Mir: Prelude to the International Space Station

In September 1993, the United States and Russian governments signed an agreement to develop, construct, and operate an orbital platform that eventually came to be known as the International Space Station (ISS). Crews would consist of both U.S. astronauts and Russian cosmonauts, with command of the station alternating between representatives of each nation from one expedition to the next.

In preparation for future ISS missions, seven NASA astronauts spent time on board Russia's Mir space station. Placed in orbit in 1986, Mir (Russian for "peace") had exceeded its planned operational lifespan but was the only existing platform capable of providing long-duration space mission experience. Since the Skylab program of the early 1970s, in which three crews had spent between 28 and 84 days in space, NASA astronauts had not experienced missions of longer than a few weeks on board the Space Shuttle. The corporate knowledge within NASA of long-duration orbital operations had been largely lost.[2]

At the time of the agreement, economic and political realities threatened both the United States and Russian space budgets, and a cooperative partnership seemed particularly attractive as the Cold War receded into memory with the 1991 dissolution of the Soviet Union. Russians needed the money and

---

2. Henry S.F. Cooper, *A House in Space* (New York: Holt, 1976); David Hitt, Owen K. Garriott, and Joseph P. Kerwin, *Homesteading Space—The Skylab Story* (Lincoln, NE: University of Nebraska Press, 2008).

support to keep Mir in orbit, and Americans needed a partner experienced in long-duration space flight. Russia's operational knowledge and scientific data in the field of systems engineering, space-station logistics, mission planning, and medical and psychosocial issues not present on shorter space missions was considered a boon by NASA. Additionally, the NASA-Mir program provided employment for many of the talented Russian space program personnel in the years before the ISS partnership finally came to fruition.[3]

There were numerous other challenges to be dealt with, however. Due to budgetary shortfalls, repairs to Mir and preventive maintenance measures were delayed. Russian Space Agency (RSA) personnel pay was hardly sufficient to meet ordinary expenses. Additionally, confusing chain-of-command issues arose because NPO Energia (a quasi-private Russian company) effectively owned Mir. A significant amount of control was also exerted by the Russian military and the RSA in a power-sharing arrangement.

The role of NASA astronauts aboard Mir was to perform a variety of scientific studies and to support the Russian crew as needed, leading to a strong division between what was considered American work and Russian work. This was illustrated dramatically during astronaut Michael Foale's tour aboard Mir in 1997, as he attempted to establish rapport with his Russian colleagues. During the first several weeks, although he concentrated on his scientific research, he felt underutilized as a member of the crew. He had noticed that his crewmates were occupied almost constantly with repairs and routine maintenance work and felt that he could easily assist with some of the more mundane tasks in order to relieve the cosmonauts' burden. Seeing the Russians falling behind in their efforts and neglecting other important items, he offered to take over some of the routine cleaning and maintenance tasks. The commander, Vasily Tsibliyev, initially rebuffed his offer, saying that the Russians were responsible for the heavy work as Foale was "a soft American poodle and not of the same kind."[4]

Sometime later, during a discussion with the Russian mission control center, Tsibliyev complained that the cosmonauts had not had a free weekend in 3 months due to their workload. Foale's repeated offer to take on some of the tasks was met with laughter from the control room, but his proposal

3. Michael R. Barratt and Sam L. Pool, *Principles of Clinical Medicine for Space Flight* (New York: Springer, 2008); Jay C. Buckey, *Space Physiology* (New York: Oxford University Press, 2006); Nick Kanas and Dietrich Manzey, *Space Psychology and Psychiatry* (El Segundo, CA: Springer and Microcosm Press, 2008); Wiley J. Larson and Linda K. Pranke, eds., *Human Spaceflight* (New York: McGraw-Hill: 2007).

4. David J. Shayler, *Disasters and Accidents in Manned Spaceflight* (Chichester, U.K.: Praxis Publishing Ltd., 2000).

was eventually accepted. A few hours later, NASA liaison controller Keith Zimmerman asked for confirmation from Foale that he had offered to do "Russian work." Tsibliyev finally began trusting the American with general station work tasks and accepted him as a true member of the crew, rather than simply a foreign researcher.[5]

There was always at least one civilian cosmonaut flight engineer on board Mir reporting to officials at NPO Energia and to the Mir commander. At the time of the Progress 234 June 1997 collision, there were three crewmembers on board the Mir station: Foale and two cosmonauts. During the previous near miss by the Progress 233 cargo ship in March 1997, the same core Russian crew was on board Mir, along with NASA astronaut Jerry Linenger. In both cases the Americans had relatively little knowledge of the manual-docking test being conducted since they were relegated to the role of observers.[6]

Vasily Tsibliyev, commander of Mir in 1997, was selected for cosmonaut training in March 1987. He made his first space flight in July 1993 as part of the Mir Expedition EO-14 crew, performing five extravehicular activities during a 196-day stay aboard the station. (NASA)

The Station Crew

Vasily Tsibliyev was a veteran cosmonaut and Russian Air Force colonel with experience as a jet pilot. Born in 1954, he was married with two children. He had graduated from Kharkov Military School of Aviation in 1975 and the Gagarin Air Force Academy in 1987. He was selected for cosmonaut training in March 1987 and made his first space flight aboard Soyuz TM-17 in July 1993 as part of the Mir Expedition EO-14 crew. During a 196-day stay on board Mir, he performed five extravehicular activities to install equipment and retrieve scientific instruments.

In preparation for departure from Mir in January 1994, Tsibliyev maneuvered the Soyuz TM-17 spacecraft around the station in order to conduct a visual inspection. While

---

5. Ibid.

6. Bryan Burrough, *Dragonfly: NASA and the Crisis Aboard Mir* (New York: Harper-Collins, 1998).

doing so, the Soyuz accidentally struck Mir at least twice at very low speed. Following a safe landing, ground processing teams discovered a number of souvenirs taken by the crew from the station that exceeded the weight limit allowed on board the Soyuz reentry capsule during landing. A team of Russian investigators suggested that the excess weight not only endangered the crew during landing but might also have contributed to problems with attitude control during the station fly-around. The most significant factor was traced to a power switch for one of two hand controllers that had been left in the off position.[7]

Similar devices were used to control the Telerobotically Operated Rendezvous Unit (TORU) manual docking system for Progress supply ships linking up with Mir. By the time of the 1997 mishap, Tsibliyev was considered one of the most skilled operators of the TORU system. He was, however, very uncomfortable with attempting manual docking maneuvers at long ranges (up to 6 kilometers). During such an attempt in March 1997, the Progress 233 spacecraft nearly collided with the station during a redocking maneuver that was a test of the TORU system. According to later testimony by Tsibliyev, the cargo vessel failed to respond to his commands. It passed within 750 feet of Mir, and future attempts to redock were abandoned in order to conserve propellants for the craft's deorbit burn. The TORU system had been designed only for short-range docking maneuvers. Longer-range docking required additional controls and displays in order to be accomplished with a margin of safety.[8]

Aleksandr "Sasha" Lazutkin served as the flight engineer for the 1997 mission. He was a civilian reporting to both the Mir commander and his superiors at NPO Energia. From 1981 to 1984, he had worked on mathematical models for thermal control systems at the Moscow Aviation Institute. In 1984 he was hired by NPO Energia to develop and optimize procedures and equipment for extravehicular activities, or spacewalks. Considered by his colleagues to be an excellent worker and easy to get along with, he had begun cosmonaut training in 1992. The Mir expedition was his first tour in space, and like many astronauts and cosmonauts, he suffered significant nausea during the early part of the mission. His main task was to assist the Mir commander as needed and

---

7. Rex D. Hall and David J. Shayler, *Soyuz: A Universal Spacecraft* (Chichester, U.K.: Praxis Publishing Ltd., 2003).

8. Stephen R. Ellis, "Collision in Space: Human Factors Elements of the MIR-Progress 234 Collision," *Ergonomics in Design* 8, no. 1 (January 2000); Dwight A. Holland, "A Systems Case Study Examination of the Near-Catastrophic Mir-Progress 234 Collision with Emphasis on the Human Factors Issues Surrounding this Mishap," Proceedings of the INCOSE Mid-Atlantic Regional Conference, 18.2-1–18.2-8, Reston, VA, April 5–8, 2000; David J. Shayler, *Disasters and Accidents in Manned Spaceflight* (Chichester, U.K.: Praxis Publishing Ltd., 2000).

Aleksandr "Sasha" Lazutkin served as the flight engineer for the 1997 mission to Mir. He was a civilian reporting to both the Mir commander and his superiors at NPO Energia. The Mir expedition was his first tour in space. (NASA)

to keep the myriad onboard systems repaired and functioning within specified parameters. This proved a daunting task on a space station where some of the components were operating well beyond their design life and showing signs of deterioration. Lazutkin was a stable family man with a wife and two daughters, and he lived in the same apartment he had occupied since 1961. He had arrived on Mir in February 1997 with Tsibliyev. He was well aware that exemplary performance on this mission would mean increased pay, which he hoped to use to move to a better home.[9]

NASA astronaut Jerry Linenger was on board Mir from January through May 1997. A goal-oriented and gifted Navy captain known for his outstanding drive and discipline, Linenger was a husband and father of a very young son, with another child on the way. He was a medical doctor and was deeply concerned about accomplishing his assigned science program on Mir. He occasionally felt the need to challenge the various Mir/NASA chains of command regarding the most effective use of his time while trying to accomplish what he believed to be the key elements of his mission. Linenger had a Ph.D. in epidemiology, in addition to several master's degrees. With this background, he was greatly concerned about the potential effects of contaminants in the station's atmosphere, about medical experiments he perceived as unwise or dangerous, and with maintaining a high degree of personal fitness. During his stay aboard Mir, Linenger became the first American to conduct a spacewalk from a foreign space station and in a non-American-made spacesuit.[10]

While living aboard the station, Linenger and his fellow crewmembers faced numerous difficulties, including the most severe fire ever to occur aboard an orbiting spacecraft; numerous failures of onboard systems (oxygen generator, carbon dioxide scrubbing, cooling line leaks, communication antenna

9. Ibid.

10. Jerry M. Linenger, *Off the Planet: Surviving Five Perilous Months Aboard the Space Station* Mir (New York: McGraw-Hill, 2000).

tracking capability, urine collection, and processing facility); the near collision with Progress 233; loss of station electrical power; and loss of attitude control, resulting in a slow, uncontrolled tumble through space. In spite of these challenges and added demands on the crew's time (made in order to carry out the repair work), all mission goals were accomplished, including all of the planned U.S. science experiments. Linenger had witnessed the near miss of the Progress 233 attempted docking but said little about it in the immediate aftermath, apparently believing that everyone in the U.S. and Russian chains of command had been fully briefed regarding the incident. He returned to Earth aboard the Space Shuttle Atlantis in May 1997, exchanging places with Foale, his replacement.[11]

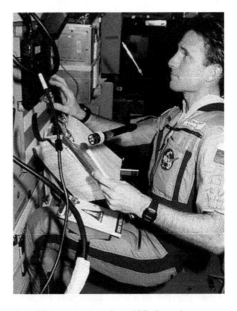

Jerry Linenger was on board Mir from January through May 1997. While living aboard the station, he and his fellow crewmembers faced numerous difficulties including fire, failures of various onboard systems, a near collision with an unpiloted cargo ship, loss of electrical power, and loss of attitude control that resulted in a slow, uncontrolled tumble through space. (NASA)

Foale, an American citizen born in 1957 to British and American parents, had an unusual background. Because his father was a British Royal Air Force Pilot, Foale spent his youth at military bases overseas and in the United Kingdom. An English boarding school education taught him how to get along with strangers. He received a bachelor of arts degree in physics and a Ph.D. in astrophysics from Cambridge University in 1982, after studying there at Queens College. While a postgraduate at Cambridge, Foale participated in the organization and execution of scientific scuba-diving projects to salvage antiquities in the Aegean Sea and the English Channel, but he was more interested in a career with the U.S. space program. He was subsequently hired by McDonnell Douglas Corporation to work on Space Shuttle navigation problems. In June 1983, Foale joined NASA at the Johnson Space Center in the payload operations area of the Mission Operations Directorate and was

---

11. Ibid.

C. Michael Foale joined the Mir crew in May 1997, arriving aboard Space Shuttle Atlantis. Foale's excellent command of the Russian language and his insistence on helping with so-called "Russian work" helped him become integrated with the crew to a greater extent than previous NASA astronauts. (NASA)

responsible for payload operations on several Shuttle missions. He had a wife and two young children.[12]

Selected as an astronaut candidate in June 1987, he initially flew the Shuttle Avionics Integration Laboratory simulator to provide verification and testing of orbiter flight software and later developed crew rescue and integrated operations for the ISS. Prior to 1997, Foale participated as a crewmember on several Space Shuttle missions, including Space Transportation System (STS)-63, the first rendezvous with the Mir space station (no docking took place on that mission). During STS-63, he made his first extravehicular activity, a 4.5-hour spacewalk to evaluate extremely cold spacesuit conditions and explore manually maneuvering the 2,800-pound Spartan satellite.

In preparation for a long-duration flight on Mir, he trained at the Cosmonaut Training Center, Star City, Russia. Foale's excellent command of the Russian language and insistence on helping with so-called Russian work helped him become integrated with the crew to a greater extent than previous NASA astronauts had. Following the collision, he helped reestablish space station operations that had been degraded when part of the station depressurized. Foale returned to Earth on October 6, 1997, having spent 145 days in space.[13]

The Spacecraft

The 1968 science fiction film *2001: A Space Odyssey* depicted an orbital space station as spacious, quiet, and spotlessly clean. Male personnel on board were dressed in business suits while women wore dresses and high-heeled shoes. The reality, embodied by Mir and the ISS, was quite different.

Mir was cramped and cluttered. It was usually hot, noisy, and smelly. Sometimes it was cold, noisy, and smelly. The crew might be attired in nothing

---

12. Clay Morgan, *Shuttle-Mir: The United States and Russia Share History's Highest Stage*, (Washington, DC: NASA SP-4225, 2001).

13. "NASA Astronaut Biographies," *http://www.jsc.nasa.gov/Bios/htmlbios/foale.html*, accessed September 5, 2010.

Unlike most space stations depicted in science fiction films, Mir was cramped and cluttered. Foale described the station as "a warm, welcoming, cozy place," despite the masses of cables, equipment, and wires that occupied every available space. (NASA)

but shorts and socks or bundled in thick thermal suits and wool caps, depending on the vagaries of the environmental control system. When Foale arrived on Mir, he felt it was "a warm, welcoming, cozy place," despite the masses of cables, equipment, and wires that occupied every available space.[14]

Mir was a 100-metric-ton vehicle in a 52-degree inclined orbit ranging between 186 and 248 miles altitude. It was composed of several modules that included the Base Block (crew sleeping/eating quarters, orbited in 1986), Kvant 1 (1987), Kvant 2 (1989), and Kristall (1990). Eventually, two more modules were added: Spektr (1995) and Priroda (1996). Crews arrived and departed via Russian Soyuz spaceships or the Space Shuttle. The Soyuz would also serve as a lifeboat in case of emergency. Mir's internal volume was roughly comparable to that of six school buses. The atmosphere on board was pressurized to an Earth-like sea level pressure.

The Soyuz and the uncrewed Progress supply spacecraft were capable of automatic rendezvous and docking with Mir when guided by the Kurs radar control system, which was produced in Ukraine, a former Soviet bloc state and now an independent republic. The Progress-M vehicle was a 7-metric-ton spacecraft capable of carrying approximately 2 metric tons of supplies to Mir. Due to

The unpiloted Progress-M vehicle was a 7-metric-ton spacecraft capable of carrying approximately 2 metric tons of supplies to Mir. It was capable of automatic rendezvous and docking with Mir when guided by the Kurs radar control system, or it could be docked using the TORU manual control system. (NASA)

disputes with regard to the disposition of the former Soviet Union's Black Sea Fleet and nuclear forces and other territorial issues, the relationship between Russia and Ukraine was often extremely strained throughout the 1990s. When the RSA asked for additional Kurs docking system support, the Ukrainian supplier demanded a price deemed too high for the financially strapped RSA. As an alternative to the Kurs system, RSA officials chose to try to extend the range of the TORU manual FCS.[15]

---

14. Morgan, *Shuttle-Mir*.

15. Burrough, *Dragonfly* (New York: Harper-Collins, 1998).

Circumstances Prior to the Collision

In March 1995, Norman Thagard (a former fighter pilot, engineer, and medical doctor) became the first U.S. astronaut to serve on Mir. Due to equipment failures, he was unable to accomplish all of his goals, and because he was not well integrated into the Russian crew, he felt somewhat underutilized and suffered some degree of boredom.

During his 4-month stay, Thagard lost a considerable amount of muscle mass (and later discovered decreased bone density, as well) despite exercise protocols to help reduce these losses.[16] These effects were not unexpected. Exercising to reduce muscle and bone loss in space is tedious, time consuming, and sometimes stressful, and it can affect hygiene issues. Additionally, running on treadmills and using other exercise equipment at certain frequencies can create mechanical resonances with the station itself that have to be mitigated. According to orthopedic surgeon Linda Shackelford's description of musculoskeletal response to space flight:

> During a typical mission of about six months['] duration, the [astronauts] lost about 1.4% overall bone mineral density, whereas the trabecular regions of the pelvis, lumbar spine, and femoral neck typically lost about 12 percent, 6 percent, and 8 percent of the initial values respectively.[17]

Shackleford also determined that of the seven NASA astronauts who flew on Mir for 4 to 6 months, "…full bone mineral density recovery occurred anywhere from six months to three years postflight for the majority of astronauts. The few who lacked full recovery in one or two regions had partial recovery in those regions, with plateau after recovery less than 5 percent below preflight values."[18] However, on the station—and more importantly, back on Earth—astronauts with weakened bones and muscles are clearly more at risk for injury. This is compounded by the fact that changing acceleration/gravity fields create neurovestibular interactions that make astronauts unsteady, dizzy, and nauseous when reintroduced to a different gravity field (an issue with major implications for a long-duration space mission, such as an expedition to Mars).

Some astronauts on lengthy space missions must also find ways to alleviate the tedium of extended exercise periods. From a human factors standpoint, perhaps virtual reality systems on future spacecraft could not only entertain

---

16. Ibid.

17. Linda Shackelford, "Musculoskeletal Response to Space Flight," in *Principles of Clinical Medicine for Spaceflight*, ed. Barratt and Pool, p. 296.

18. Ibid., p. 297.

spacefarers, but also help astronauts to simulate running in different locales, or favorite places, during extended exercise sessions.[19] There also has been discussion of using virtual reality imagery not only to help astronauts refresh critical procedural and cognitive skills as a training aid and learn new ones as required, but also to aid in preparation of their neurovestibular systems for upcoming changing acceleration and/or gravity environments.[20]

Other NASA astronauts assigned to Mir included Shannon Lucid and John Blaha, as well as Linenger and Foale. Each became integrated to varying degrees with their Russian crewmates despite a language barrier and difficult working conditions. Whereas Thagard had felt undertasked, Blaha at times felt so overscheduled by his NASA managers that it simply wore him out; his Russian crewmates suggested he slow down and get more rest. Blaha reported sleeping at times only 3 hours per night for extended periods, and he often felt isolated and alone due to cultural and language barrier issues. He also had a tense relationship with Mir Commander Valeri Korzun throughout most of his stay, which exacerbated the problem. Blaha felt that Korzun micromanaged him and did not trust him to use good judgment. As a veteran of multiple Shuttle missions, Blaha resented this treatment. Korzun apparently was afraid his "scorecard" results from Russian mission control authorities at the end of the mission would be tainted by any mistakes, and he did not want the financial and personal backlash resulting from any American astronaut's inadvertent mistakes. One of Blaha's comments to Linenger during crew transfer was that Linenger should not take the Mir commander's actions personally. To his credit, Korzun seemed to have learned from his difficult interactions with Blaha, and he got along much better with Linenger.[21]

Blaha was so physically deconditioned upon returning to Earth that he could hardly move to leave the Space Shuttle. Such physical effects are of great concern to NASA scientists planning future long-duration space expeditions. Consequently, they have studied ways to reduce the effects of microgravity

---

19. Mark A. Guidi and Dwight A. Holland, "Operational and Human Factors Implications of Physiological Deconditioning in Long Duration Spaceflight," in *Aerospace Systems*, Human Factors and Ergonomics Society Annual Meeting Proceedings, 1992, pp. 116–120.

20. Dwight A. Holland and William R. Barfield, "Some Virtual Reality and Telemedicine Applications Useful for Long Duration Spaceflight from a Systems Engineering Perspective," in *Medicine Meets Virtual Reality*, ed. Westwood, Hoffman, Robb, Stredney (Washington, DC: IOS Press,1999), pp. 141–147; Donald E. Parker and K.L. Parker, "Adaptation to the Simulated Stimulus Arrangement of Weightlessness," in *Motion and Space Sickness*, ed. G. Crampton (Boca Raton, FL: CRC Press, 1990), pp. 247–262.

21. Wiley J. Larson and Linda K. Pranke, eds., *Human Spaceflight* (New York: McGraw-Hill, 2007).

on various organ systems, develop protection against cosmic radiation, and improve psychological states and group dynamics.[22]

To varying degrees, long-duration space travelers may suffer from asthenia. This condition may include any of the following symptoms: psychological or emotional fatigue or weakness, hypoactivity, irritability and tension, emotional instability, appetite and sleeping problems, attention and memory deficits, emotional withdrawal from others, and territorial behavior.[23]

Similar, and other, psychological and group-dynamics problems have been reported on research teams in the Antarctic, on submarines and ships, and in other isolated and confined environments.[24] These psychological, cultural, and group-dynamics issues have been reported and summarized in a wide variety of literature over many years.[25]

One psychiatrist–flight surgeon has suggested that since most cosmonauts were selected from similar military and flight-test backgrounds, they could be expected to get along quite well during extended missions simply due to shared background and common mission goals.[26] Research into the experiences of teams working in isolated, confined environments such as polar outposts has demonstrated, however, that this is not typically the case. Nevertheless, this general sentiment seemed to underlie the scheduling of tasks on board Mir, the competition for spacecraft resources, and other factors, inadvertently putting the spacefarers on board Mir at odds with each other on longer flights. On a space station, it is not possible to simply take a walk outside to disengage from a stressful situation. Crewmembers were constrained to seeing the same faces and personalities day after day, week after week for the duration of the mission.

---

22. Barratt and Pool, *Principles of Clinical Medicine for Space Flight*; Buckey, *Space Physiology*; Kanas and Manzey, *Space Psychology and Psychiatry*.

23. Kanas and Manzey, *Space Psychology and Psychiatry*.

24. Jack Stuster, *Bold Endeavors: Lessons from Polar and Space Exploration* (Annapolis, MD: Naval Institute Press, 1996).

25. For examples, see B.J. Bluth, "The Benefits and Dilemmas of an International Space Station," *Acta Astronautica* 11 (1984): 149–153; Andrew Chaikin, "The Loneliness of the Long-Distance Astronaut," *Discover* (February 1985): 20–31; Albert A. Harrison, Y.A. Clearwater, and Christopher P. McKay, eds., *From Antarctica to Outer Space* (New York: Springer, 1991); Robert L. Helmreich, "Culture and Error in Space: Implications from Analog Environments," *Aviation, Space, and Environmental Medicine* 71, no. 9, section 2, suppl. (September 2000), A133–A139; Dwight A. Holland, "Systems and Human Factors Concerns During Long-Duration Spaceflight" (M.S. thesis, Virginia Polytechnic Institute and State University, Blacksburg, VA, 1991); Kanas and Manzey, *Space Psychology and Psychiatry*.

26. P. Santy, "The Journey In and Out: Psychiatry and Space Exploration," *American Journal of Psychiatry* 140 (1983): 519–527.

By the 1990s, NASA managers were beginning to learn that overmanaging astronauts could be detrimental during long-duration space missions.[27] Space Shuttle missions tended to be relatively short (no more than a few weeks), and the schedules could be managed tightly; but for longer missions, this approach is not productive. In fact, similar issues had arisen during the NASA Skylab space station missions of the 1970s, when astronauts had complained about tight, inflexible timelines.[28]

Chronic micromanagement and overscheduling also led to frustration, burnout, and fatigue in mission control. A study noted lessons learned from the NASA-Mir program:

> Phase I lessons have emphasized that astronaut training objectives for long-duration crew members will differ from those NASA has traditionally employed for Shuttle crews. It is essential to address psychological factors early to maintain crew morale and efficiency throughout long-duration stays. Overall, mission training must be more general-skills-oriented than the intensive procedural practices that are emphasized in Shuttle training. Skills training will provide better flexibility and is more cost-effective for on-orbit station operations.
>
> In conjunction with this emphasis on skills training, NASA will schedule on-orbit crew activities for the ISS very differently than the way it does for Space Shuttle missions. Shuttle missions are planned in great detail before flight to make optimum use of every available moment. Station crews will perform a wide range of duties, both planned and unplanned, and the Phase I experience has taught us that it is neither practical nor feasible to create extremely detailed day-to-day timelines for long-duration space station operations. The Russian program uses a more flexible approach to scheduling, in which crewmembers apply the fundamental skills they learned in training to the tasks required by the actual priorities of the day. For instance, solving a problem with an experiment's equipment may require sending a replacement part or repair tool to the station on an interim

27. JSC, *Off the Planet; Lessons Learned on Skylab Program* (Houston, TX: NASA TM X-72920, 1974).

28. Henry S.F. Cooper, *A House in Space* (New York: Holt, 1976); David Hitt, Owen K. Garriott, and Joseph P. Kerwin, *Homesteading Space—The Skylab Story* (Lincoln, NE: University of Nebraska Press, 2008); William R. Pogue, *How Do You Go to the Bathroom in Space?* (New York: Tom Doherty Associates, 1985).

flight. Meanwhile, rearranging the research agenda would free time later to complete the problematic experiment once the repairs are complete.

This approach ensures that long-term goals for the entire mission are met. Phase I has verified ISS program planning and timelining tools; modifications and enhancements are being made to ISS flight planning strategies and concepts on the basis of this experience.[29]

Whereas long-duration space flight missions are like marathons, Shuttle missions have been like sprint races. The NASA-Mir missions served a useful purpose for later ISS operations by allowing mission planners to discover the strengths and weaknesses in U.S. and Russian management styles, logistics concepts, and hardware issues. This helped avert later problems.

Following Linenger's arrival on Mir in January of 1997, he attributed some of his difficulties to an accelerated course in Russian that left him less than fluent in the language. For a variety of reasons, he had also experienced conflicts with some of the Russian training staff about issues he felt were fundamental. As a result of this, and his own admitted tendency to be egocentric, the RSA psychological support staff recommended that he not be allowed to fly. They simply felt he would not be able to fit in well enough with the rest of the crew. Fortunately, with support from his flight surgeon in Moscow and his NASA flight surgeon, along with high-level NASA management, he was cleared to fly.[30]

After Linenger reported on board Mir via a Shuttle flight and received his days-long transition briefing by Blaha, he settled into a routine of work and exercise. He immediately noticed that space-to-ground communications were often short and of poor quality and that the Russians seemed unwilling give him the time that he needed to confer with his NASA liaison about various matters for which he was responsible on the station. This had frustrated previous astronauts as well and slowly infuriated Linenger. He saw good communications with the ground as essential for the orderly conduct of station work and daily functioning in a safe environment. He was also frustrated at having to wait to talk to his wife while the Russians used the radio to discuss seemingly inconsequential matters.

29. NASA, summary of several documents on lessons learned for Phase I International Space Station, NASA Human Space Flight, *http://spaceflight.nasa.gov/history/shuttle-mir/history/h-b-lessons.htm*, uploaded 1998, accessed September 3, 2010.

30. Linenger, *Off the Planet*.

Mir was plagued with many small problems, such as antifreeze-filled cooling pipes that leaked due to corrosion and had to be repaired using flexible tubes as seen here, held by Jerry Linenger. Goggles and filter masks were necessary due to floating blobs of antifreeze, metal shavings, and other particulate debris. (NASA)

Conditions On Board Mir

In February 1997, there were multiple systems problems on Mir that kept the crew busy. Between electrical and power problems, attitude control issues, ethylene glycol leaks, fumes in the atmosphere, and overtaxed carbon dioxide scrubbers, the crew was busy with basic maintenance of the aging station. The oxygen-generating system was unable to keep pace with the crew's consumption, particularly when six people were on board, as during crew changes (Blaha leaving, Linenger arriving). Linenger was eventually told to refrain from exercise for a brief period in order to conserve oxygen and reduce carbon dioxide levels. Linenger, however, was reluctant to give up his regimen, which was important for his cardiac, muscle, and bone health (as well as for his mental health, since he was an athlete and enjoyed exercise for stress reduction).

A serious incident occurred in February 1997 in which a crewmember activated a lithium perchlorate candle (an oxygen-generating canister) that caught fire, shooting a blowtorch-like flame as much as 3 feet long. Thick, noxious smoke instantly filled the interior of the module. The crew fought the blaze while considering the possibility that it might become necessary to abandon the station, but two additional problems surfaced that would have complicated egress and departure. First, the flame blocked access to one of two Soyuz craft docked to Mir. Second, there was only a single copy of the Soyuz retroburn schedule needed for reentry, and a second had to be printed from the station computer. The fire was eventually extinguished, but the crew had to wear respirator masks in the polluted cabin atmosphere. The first respirator that Linenger tried to use proved useless. Additionally, the first fire extinguishers used had no effect, and the canister may have simply burned itself out. Although the fire had only burned for about 90 seconds, Linenger later said it had felt like 15 minutes.[31]

Korzun tried to play down the seriousness of the incident when describing it to mission control. Linenger felt the cosmonaut was underplaying the seriousness of the fire for public relations purposes (RSA needed the revenue from countries willing to send various individuals to Mir). On Earth, NASA issued the following public statement: "A problem with an oxygen-generating device...set off fire alarms and caused minor damage to some hardware on the station...a small fire burned for about 90 seconds...and was easily extinguished by the crew."[32] After the fire, Linenger checked the crew's blood oxygen saturation levels and discovered no apparent lingering respiratory effects.

In late February, Linenger reported difficulty sleeping for several nights due to continuing poor atmospheric conditions and having to wear the respirator masks. The interior of the station was also growing hot and humid due to the overtaxed environmental control systems. Linenger noted that the ambient temperature was around 90 degrees Fahrenheit and that the high humidity encouraged the growth of mold and bacteria in the station.[33]

After Korzun and another cosmonaut, Aleksandr Kaleri, departed Mir, only Linenger, Tsibliyev, and Lazutkin remained. Linenger noted the sometimes-strained relationship between the Mir crew and mission control:

> The Russian experts saw me as egocentric, independent, a rebel— in short, a typical American. They hear Vasily's temper flare on numerous occasions when he disagreed with the actions that the ground advised and, especially near the end of our flight together, they heard him rant and rave about how messed up things were on MIR and how it was never this bad during his first tour of duty on the space station. Vasily privately attributed the MIR's problems to the economic decline in Russia.... As a result MIR was, in Vasily's words, "falling apart." Not wanting to indict themselves, the people on the ground control concluded that the American onboard caused Vasily's anger.
>
> While my interactions with both MIR crews were almost unbelievably positive, the relationship between the MIR-23

---

31. Shayler, *Disasters and Accidents in Manned Spaceflight.*
32. Linenger, *Off the Planet.*
33. Ibid.

crew (Tsibliev, Lazutkin, and me) and the Russian flight controllers at mission control in Moscow was unexpectedly dismal and extremely tense.

Given my medical background in the study of human psychology, in which I specialized to some extent on the problems and adaptive strategies of people living in isolation, I was both astonished and appalled at how poorly the Russians, who had more than eleven years of MIR experience, handled the psychological aspects of long-duration flight. Mission control in Moscow became our enemy rather than our friend, our nemesis, rather than our support structure. During our time aboard MIR, mission control in Moscow repeatedly offered us calculated misrepresentations of facts. In serious situations, they deliberately omitted information until, by the end of the mission, we had no confidence in them. Nor did we feel that we could trust anything they told us.[34]

The 1997 fire was not the first time an oxygen canister had ignited. During a mission in October 1994, a canister caught fire, but cosmonaut Valery Polyakov had managed to smother it with an extra uniform.[35]

Problems aboard the station continued to mount, putting further strain on the crew. Communications with the ground became increasingly difficult. In early March 1997, Tsibliyev's attempt to practice a manual docking of the Progress 233 cargo ship resulted in near-collision, and the Mir commander became increasingly frustrated with mission controllers.[36]

In early March, the remaining oxygen-generating unit failed, necessitating replacement with a backup. The cooling system failed for several weeks, droplets of ethylene glycol leaked from a coolant loop into the station's atmosphere, and the temperature soared above 90 degrees Fahrenheit. A cycle of repairs and failures wore crew morale down little by little through fatigue and stress. In June, there were more coolant leaks. Tsibliyev felt ill for days after inadvertently floating though a large blob of ethylene glycol on June 5. The beginning of a sleep study 7 days later severely interrupted crew rest, and Tsibliyev suffered from acute/chronic sleep deprivation.[37]

---

34. Ibid.

35. Ibid.

36. Burrough, *Dragonfly*; Jeffrey Kluger, "How I Survived Mir: A Bad Day in Space," *Time* (November 3, 1997): 84–91.

37. Ibid.

Collision in Space

Despite the nearly disastrous results of the Progress 233 docking test, Russian ground controllers asked Tsibliyev to repeat the attempt with Progress 234, with the objective of determining the long-range capabilities of the TORU manual docking system. The Mir commander was ordered to shut down the Kurs radar to avoid potential interference, but this deprived the cosmonaut of critical range data. Controllers also insisted that he limit fuel usage, conserving propellants for later use.[38]

The Mir commander was uncomfortable with the planned maneuver but made a concerted effort to prepare for it. He spent several days practicing with

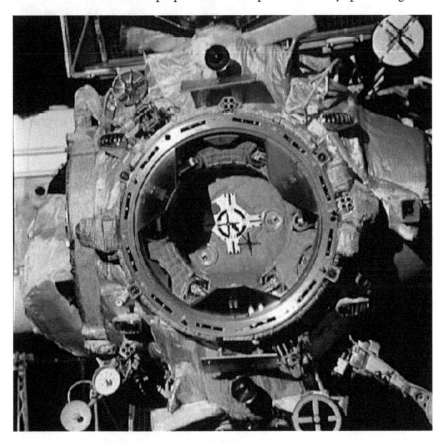

The androgynous peripheral docking system was designed to achieve the capture, dynamic attenuation, alignment, and hard docking of spacecraft through the use of essentially identical docking mechanisms attached to each vehicle. A black-and-white target served as a visual reference during docking maneuvers. (NASA)

---

38. Ellis, "Collision in Space: Human Factors Elements of the MIR-Progress 234 Collision."

Tsibliyev was considered one of the most skilled operators of the TORU system that remotely controlled the Progress cargo ships. He was, however, very uncomfortable with attempting manual docking maneuvers at long ranges (up to 6 kilometers). (NASA)

the TORU system, but his crewmates noticed signs of tension. Foale received no briefing on the test and assumed that the maneuver had been well planned by experts on the ground.[39]

On June 25, Tsibliyev allowed Progress 234 to drift away from Mir and took manual control of the cargo ship at a range of 6 kilometers from the station. To improve safety margins, the initial engine burn to bring the uncrewed craft back to Mir was designed to target a point approximately 1,000 meters behind the station. Progress 234 was given a 400-meter displacement out of the station's orbital plane, complicating the orbital dynamics of the trajectory. Tsibliyev had originally planned to establish a 5.0-meter-per-second rate of closure, but Progress 234 actually approached Mir at a rate of 6.5 meters per second.[40]

Tsibliyev aligned the cargo ship to acquire an image of Mir in the TORU system's video screen and centered Progress 234 for approach. Using a dual-joystick hand controller, he maneuvered the cargo ship, braking as necessary, using a visual reference on a television monitor that showed the station as it would appear from the Progress vehicle. Because the Kurs radar was not active,

---

39. Shayler, *Disasters and Accidents in Manned Spaceflight.*

40. *Seventh Report of the NASA Advisory Council Task Force on the Shuttle-Mir Rendezvous and Docking Missions* (Washington, DC: NASA, 1997–1999).

The collision damaged the station's solar power array, buckled a radiator, and punched a hole in the Spektr module. Only the quick reaction by the crew prevented a tragedy. (NASA)

the commander's only source of range-rate data was the changing angular size and position of Mir on the monitor. Because of the poor quality of the imagery, he failed to realize that his closure rate was too high. At a range of approximately 250 meters (about 1 minute before impact), he began continuous braking and lateral (downward) translation to avoid collision.[41]

The effort was not sufficient. With a loud thump, Progress 234 struck the station's solar arrays at about 3 meters per second. As the craft bounced off the outer hull of Mir's Spektr science module, the crew felt a sudden change in air pressure followed by the sound of warning sirens and, worst of all, precious air hissing into the void of space.

In order to seal the Spektr section from the other modules, the crew had to disconnect cables and hoses that snaked through the hatchway. There was no way to remove all of them quickly, so flight engineer Lazutkin used a knife to cut the last few, with live wires showering sparks. Low internal pressure prevented closure of the Spektr hatch because it was hinged in the wrong direction, so Lazutkin and Foale grabbed an external hatch cover. Fortunately, the pressure differential created enough suction to seal the opening.[42]

---

41. Ellis, "Collision in Space."

42. Shayler, *Disasters and Accidents in Manned Spaceflight.*

The force of the collision caused the station to begin tumbling, moving the solar panels out of alignment with the Sun. This resulted in a severe reduction in power. The station's power supply was already compromised because of the severed cables. Eventually, station attitude control was restored.

Cycle of Breakdowns

Many factors contributed to this mishap, not the least of which was physical and psychological tolls imposed on the crew by the constant cycle of systems breakdowns. One area of interest is the fact that lessons from the earlier Progress 233 incident were not applied to the Progress 234 docking attempt. It is likely that Foale was not fully apprised of the previous incident by his predecessor because Linenger was simply worn down by the various challenges he and the others had faced on Mir. As a result, the corporate knowledge of the Progress 233 incident was not properly assessed, categorized, and retained in time to influence the Progress 234 test. All decisions regarding the test rested with the Russians. As an outsider, Foale would have had no influence on his own, but a full and accurate assessment of the situation with reference to the March near-miss may have enabled NASA officials to exert some influence over Russian decision making as to the advisability of a second docking attempt.

This highlights an area of concern in accident analysis known as enabling factors—which may affect group dynamics and organizational control issues—and hints at problems with overall group/organizational situational awareness. It also specifically relates to individual situational awareness on Tsibliyev's part during the docking maneuver.[43]

Neither the NASA officials on the ground nor the astronaut crewmembers on board Mir truly understood the technical factors involved in attempting a manual docking of the Progress ship. The NASA team members simply lacked the necessary level of situational awareness required by the docking attempt attributes.[44] Furthermore, it appears that officials at NASA and the RSA were not communicating effectively in regard to the nature and risks of the docking attempts. Why the RSA apparently did not inform NASA of the details of such

---

43. Dwight A. Holland, "Systems and Human Factors Concerns During Long-Duration Spaceflight" (M.S. thesis, Virginia Polytechnic Institute and State University, Blacksburg, VA, 1991); Holland, "A Systems Case Study Examination of the Near-Catastrophic Mir-Progress 234 Collision with Emphasis on the Human Factors Issues Surrounding This Mishap," *Proceedings of the INCOSE Mid-Atlantic Regional Conference*, Reston, VA, April 5–8, 2000.

44. Eduardo Salas, C. Prince, David P. Baker, and L.B. Shrestha, "Situation Awareness in Team Performance: Implications for Measurement and Training," *Human Factors* 37, Santa Monica, CA (1995): 123–136.

an important test is unclear, given the previous 233 near-accident. It is likely the Russians were so busy with the many challenges they faced prior to the second attempt that they simply did not have the time and personnel resources needed to properly consult NASA about the upcoming June docking test. Alternatively, such behavior may have been characteristic of Russian culture, history, and practices for long-duration operations, or simply a desire to protect their public image. In fact, due to economic conditions in Russia at the time, the RSA had fewer resources to spare for the purpose of briefing NASA personnel on Mir operations that fell under the Russian purview. Despite any of these factors, it has to be considered a systemic failure as well as a less-than-optimal judgment to choose not to communicate with a partner about such an issue.

Cultural Issues

Cultural organizational/supervisory issues contributed to this mishap and are key to understanding why a highly skilled cosmonaut like Tsibliyev would attempt such a hazardous maneuver given the previously unsuccessful trial. Traditionally, the Russian space program has had a high degree of military control and oversight. The Mir commander was a Russian Air Force colonel who was accustomed to operating within a system where individual initiative was often suppressed. Russian flying tactics historically have depended heavily upon ground-control radar guidance and on a top-down approach to flying operations, even for fighter pilots. Another cultural factor affecting Russian space operations is that cosmonauts historically have been in a "worker bee" type of mode with respect to ground control.[45] In this case, mission control designated items on a "scorecard" as important to complete. If a cosmonaut did not complete a task or set of tasks, then ground control would assign a "black mark" to the cosmonaut, and that cosmonaut would receive less money at the end of the mission. Thus, mission control was judge and jury in an incentive- and punishment-based reward system for assessing cosmonaut performance.[46]

Cosmonauts were thus not motivated to challenge questionable task assignments, since doing so might result in administrative (and financial) punishment. At a time when Russia's economy was undergoing major transition, these cosmonauts depended upon their performance-based salaries. Such high-level predisposing factors for systems performance may explain why Tsibliyev apparently was willing to make the second docking attempt: he was simply following orders.[47]

---

45. James E. Oberg, "Shuttle-Mir's Lessons for the International Space Station," *IEEE Spectrum* (June 1998): 28–37.

46. Holland, "A Systems Case Study Examination of the Near-Catastrophic Mir-Progress 234 Collision."

47. Ibid.

The Mir commander was considered one of the most skilled operators of the TORU manual docking control system in the cosmonaut corps. Although he had protested attempting a second long-range docking, his flight engineer tried to encourage him. Tsibliyev replied, "It's bad.... It's a dangerous thing to do."[48]

As Mir commander, he reported to both military and civilian authorities through a complex chain of command. Several parties had stakes in the decisions that were made: the Russian military, the management on the ground (a mixture of Russian space officials and representatives of NPO Energia), and the ground controllers themselves.

There were a variety of complex approach rules that were supposed to be followed at the direction of ground control personnel if a given system component was malfunctioning. In the case of the Progress 233 incident, one Russian flight controller, claiming that Tsibliyev had not properly followed the designated approach rules, broke off the attempt.[49] An investigation showed that either the TORU camera had failed for some reason (possibly interference from the Kurs radar), or Tsibliyev forgot to press the camera switch to turn it on (or that he did not depress the switch with enough force to engage it). The system apparently did not have a fault-detection system that enabled the operator to assess whether the camera was activated and functioning properly, and because communications with the ground were intermittent, ground control was unable to monitor his attempt to dock Progress 233 in real time as a backup.[50]

Problems in these areas relate to "preconditions for unsafe acts" and "unsafe acts," as defined by the Human Factors Analysis and Classification System (HFACS). They are defined as contextual and person-machine communication issues.

Fatigue Development

Tsibliyev's months of interrupted sleep cycles undoubtedly contributed to the accident. High workload combined with a host of small emergencies led to generalized chronic fatigue weeks before the decision was made to attempt the Progress 234 docking test. If this general "macro-task environment" were not challenging enough for the crew, a 2-week-long sleep study that began on June 12 had a further negative effect on the crew's sleep quality. This study added another acute fatigue element to an already bad chronic fatigue situation. Fatigue is widely known to affect interpersonal relations, decision making,

---

48. Kluger, "How I Survived Mir."

49. Public Broadcasting System, "Terror in Space," *NOVA*, transcript no. 2513 (Boston, MA: Public Broadcasting System, 1998).

50. Burrough, *Dragonfly*.

and the ability to carry out psychomotor tasks requiring higher-level cognitive functioning.[51] In fact, based upon a wide variety of aviation studies and recent noncombat U.S. Air Force multicrew aircraft mishap investigations, fatigue alone is thought to be capable of substantially reducing a crew's situational awareness, judgment, and ability to perform effectively.[52]

It is therefore not surprising that the Mir commander, given his chronically fatigued state, was overextended physically and mentally by the time of the Progress 234 docking attempt.[53] Furthermore, there is evidence that discussions were held within the Russian space program as to whether a docking should be attempted at all in light of the myriad factors involved in the task, including fatigue.[54] Russian space officials ultimately decided that Tsibliyev should proceed despite his objections. This category of analysis falls into the preconditions for unsafe acts region in all three main categories of environmental factors, condition of operator, and personal/interpersonal factors, in addition to supervisory errors.

From a human factors perspective, the Mir commander's situational awareness was degraded dramatically during the manual Progress 234 docking attempt due to the inadequate visual display and control information, particularly with regard to the critical parameters needed, such as the position, range, and velocity information of the approaching Progress 234 supply ship. Predictive displays to help control the Progress/Mir approach trajectory would have been key in this scenario to avoiding a mishap.

---

51. Mary Conners, Albert Harrison, and Faren Akins, *Living Aloft–Human Requirements for Long-Duration Spaceflight* (Washington, DC: NASA SP-483, 1985).

52. H. Van Dongen, "Comparison of Mathematical Model Predictions to Experimental Data of Fatigue and Performance," in Proceedings of the Fatigue and Performance Modeling Workshop, June 13–14, 2002, Seattle, WA, ed. Neri and Nunneley, *Aviation, Space, and Environmental Medicine* 75, no. 3, section II (2004): A15–36.

53. Jeffrey J. Armentrout, Dwight A. Holland, Kevin J. O'Toole, and William R. Ercoline, "Fatigue and Related Human Factors in the Near Crash of a Large Military Aircraft," *Aviation, Space, and Environmental Medicine* 77, no. 9 (2006): 963–970; Michael B. Russo, Helen C. Sing, Saul Santiago, Athena P. Kendall, Dagny E. Johnson, David R. Thorne, Sandra M. Escolas, Dwight A. Holland, Stanley W. Hall, and Daniel P. Redmond, "Occurrence and Patterns of Visual Neglect in U.S. Air Force Pilots in a Simulated Overnight Flight," *Aviation, Space, and Environmental Medicine* 75, no. 4, section I (2004): 323–332; Michael B. Russo, Athena P. Kendall, Dagny E. Johnson, Helen C. Sing, Sandra M. Escolas, Saul Santiago, Dwight A. Holland, Stanley W. Hall, and Daniel P. Redmond, "Visual Perception, Psychomotor Performance, and Complex Motor Performance During an Overnight Air Refueling Simulated Flight," *Aviation, Space, and Environmental Medicine* 76, no. 7, section II (2005): 92–103.

54. Burrough, *Dragonfly*.

These remarks illustrate the HFACS model component known as preconditions for unsafe acts, as well as the unsafe act itself (in this case, the docking maneuver). The Mir-Progress collision incident included a variety of preconditions for unsafe acts. These can be categorized as environmental factors (heat and humidity, inadequate controls/displays), condition of the operators (the crew was tired mentally and physically), and interpersonal factors (coordination and communication among crew and outside were ineffective, and the crew was less than fit for duty). There were also questions raised about violations of mission rules.[55]

Self-Protection and Currency of Training

As noted, Tsibliyev was regarded as one of the best pilots in the cosmonaut corps and was considered a skillful operator of the TORU manual docking system. From a human factors perspective, the Progress 234 docking attempt was problematic, first, because of Tsibliyev's lack of recent training prior to the mishap. Additionally, Mir lacked adequate simulation facilities on which he could practice such maneuvers. From the human factors standpoint regarding person-machine interface training, this was a critical concern. Without such practice and training, Tsibliyev's situational awareness for the task at hand was substantially degraded, leading to problems in judgment and decision making.

Immediately following the impact, the crew was forced to spend an inordinate amount of time trying to seal the Spektr module to mitigate the hull breach. It reportedly took approximately 11 minutes to secure the Spektr module, delaying actions to stabilize the tumbling station. The overall air pressure within Mir dropped from 760 millimeters of mercury (mmHg) (sea-level equivalent) to around the 670 mmHg. The minimum allowable pressure limit permitted before abandoning the station was 550 mmHg, and it was estimated that the crew had about 28 minutes to evaluate and stabilize the station's condition. If the crew had not managed to seal off the damaged module in time, Mir would have been abandoned to plunge, out of control, into Earth's atmosphere, potentially falling on a populated area. NASA engineering directives for U.S. space stations require that there be quick-disconnect capability for cables running between modules that may have to be isolated in an emergency.[56]

Crew Ambient Physical Environmental Concerns

Constant monitoring of a space station's atmosphere is required to keep the correct balance of oxygen, nitrogen, and carbon dioxide within tolerable limits.

---

55. Holland, "A Systems Case Study Examination of the Near-Catastrophic Mir-Progress 234 Collision."
56. Ibid.

This was particularly true for the levels of oxygen and carbon dioxide on board Mir. One of the two oxygen generators on Mir often seemed to be malfunctioning, requiring the use of lithium perchlorate canisters to produce additional oxygen. The carbon dioxide scrubbers and various oxygen generators were pushed to their limits throughout the mission. This situation had human factors implications since crewmembers were unable to exercise as often as they should have to combat the deconditioning effects of microgravity, release stress, and help maintain a feeling of well-being. Such basic factors as exercise and well-being are fundamental for long-duration space missions in a microgravity environment. The hot and humid environment added to the stressful conditions, making it more difficult to work and sleep, creating mental states less capable of focusing and adapting, raising tensions, decreasing the general condition of the operators, and generally increasing the likelihood of error.

The Human-Machine Interface

The control-display interface and the management decision that urged the Mir commander to proceed with the extended manual rendezvous were the most critical contributors to the mishap. Just before the Progress 234 docking attempt, Tsibliyev was advised that fuel should be conserved to the extent possible since the Progress ship was going to be used after the docking attempt in a flyaround inspection mode. This may have constrained the Mir commander's willingness to maneuver Progress earlier in the approach and in a more proactive manner. Additionally, the Progress 234 supply ship initially was assigned to a 400-meter displacement out of Mir's orbital plane, complicating the dynamics of the approach trajectory. Orbital mechanics and rendezvous control between two flying spacecraft are notoriously counterintuitive. Problems faced by Tsibliyev included an inadequate technological environment (control interface), failure to make timely decisions (situational awareness), skill errors (training problems), and perceptual errors.

During the docking attempt, the Progress 234 approached Mir too quickly—6.5 meters per second (versus a planned 5.0-meters-per-second velocity). The Kurs radar was shut off due to concerns that it had interfered with TV video feed problems on the previous attempt (technology/interface and perceptual errors). Tsibliyev therefore did not have reliable range and velocity data, as was usually the case when the Kurs radar ensemble was activated. A hand-held laser range finder and clock were also used to try to assess the approaching Progress's range and velocity (technological environment issue). Unfortunately, the Progress 234 was out of sight for much of the approach, so accurate range/velocity information was not available (a general Precondition for Unsafe Acts issue). Members of the crew frantically tried to find the oncoming Progress craft by looking out of Mir's windows to obtain

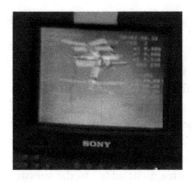

This television screen provided the only direct visual data to the TORU operator during manual docking maneuvers. During the failed docking attempt, Mir appeared on the monitor as a pixelated and blurry image that seemed to fade in and out of view. (NASA)

laser range measurements. Flying in this mode of operation required that a reasonable line of sight be maintained throughout the Progress 234's approach to Mir.

Because the Progress ship was approaching Mir from a plane appearing to the observer to be above Earth, the Progress camera was looking down at Mir as it orbited Earth. Given the poor Progress camera/Sony monitor resolution and the poor visual contrast of Mir against the clouds below on that particular day, it was difficult at best for Tsibliyev to perceive and estimate the rate at which the Progress ship was approaching the station. This, again, led to perceptual errors. Mir appeared on his TV monitor as a "noisy," pixilated, and blurry image that seemed to fade in and out of view as the clouds rolled by below. There was insufficient contrast between Mir and the clouds, along with the poor resolution of the picture, to enable use of the grid technique superimposed on the visual frame for accurate velocity estimation (control-display interface/technology problem). This graphical method for range/velocity estimation is based upon the known sizes of certain structures as they appear on a grid overlaid on the Sony video monitor. When Progress was only about 250 meters from Mir, the crew finally realized that the cargo ship was approaching too fast. Tsibliyev applied braking and maneuvering thrusters immediately, but it was too late. By then, Progress 234 had too much momentum to be turned quickly with the available thrusters.[57] At best, Tsibliyev might have been expected to detect 5-percent changes in speed. But because the viewing conditions on the TV monitor were degraded, he would have been much less sensitive to changes in the range to Mir and to relative velocity.[58]

Tsibliyev developed heart arrhythmias following the collision, in addition to other minor medical complaints, and needed anxiolytics. This may have been due to concern that he would be blamed for the accident and his performance called into question. Such concerns are strongly embedded in Russian organizational culture.

---

57. Ibid.

58. Ellis, "Collision in Space."

Final Thoughts

The events surrounding the Mir-Progress docking incidents highlight, from a systems theory and human factors perspective, the various causal factors that combined to result in a nearly disastrous collision.

The lessons learned may be readily applied to future space flight operations if fundamental space systems/human factors and organizational engineering principles are applied. If long-duration space missions can be designed and thought of in terms of accident prevention (and how system attributes interrelate and may feed into a multicausal accident sequence), then perhaps more safe and efficient human-rated space operations may be expected in the future.

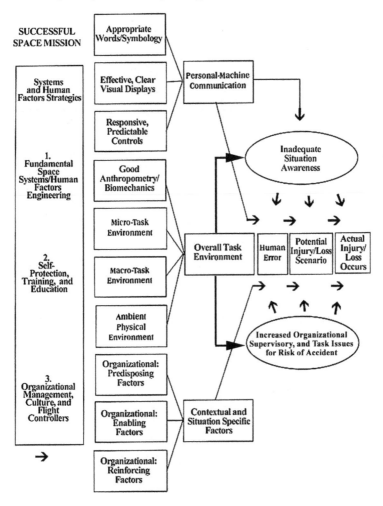

This chart illustrates basic space systems and human factors concerns for long-duration space flight in terms of mishap prevention. (Author's collection)

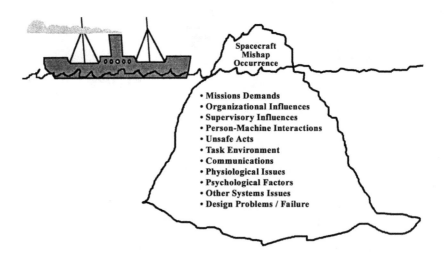

Lurking beneath the surface like an iceberg, a number of direct and indirect human factors components contributed to the Mir-Progress mishap. (Author's collection)

Conclusions

This collection of case studies highlights human factors lessons learned from a variety of aviation and space mishaps involving vehicle design, human physiology, and organizational issues. The authors examined each case from the perspective of how human factors interact with mechanical systems and human organizations, supporting a multilevel analysis in which the accident is just the tip of an iceberg underlying human-system integration efforts. The themes highlighted in the incidents described emphasize the need for attention to human factors engineering in the planning and execution of test and operational systems.

Statistics indicate the significance of human factors in aerospace-related mishaps. Information from the Naval Safety Center collected between 1990 and 2008 reveals that about 90 percent of Navy mishaps had human factors issues as a contributing cause of accidents during that period. Historically, about 80 percent of Air Force mishaps have included human factors and aeromedical issues as definite or probable contributing factors.[1]

Human factors engineering should be applied in the design phase of any aerospace vehicle. The cockpit is the primary human-machine interface. As some of the preceding case studies have shown, improvements in controls and displays likely would have prevented some accidents.

Physiological factors are an ever-present concern, particularly in high-performance vehicles that push crewmembers to their physical limits. Phenomena such as G-LOC and A-LOC will continue to pose significant problems as aircraft designers increase the speed and maneuverability of advanced combat aircraft. The development of an effective system to predict the onset of g-induced cognitive impairment—such as using near-infrared spectroscopy to monitor the brain for sudden drops in blood volume and oxygen—would save lives.

Organizational issues may be the most pervasive problem of all. Procedural problems, inadequate supervision, improper crew resource management, and lack of communication can lead to disaster. The Human Factors Analysis and

---

1. Nick Davenport, "Let's Keep You Flying," *Navy and Marine Corps Aviation Safety Magazine* 55 (January–February 2010): 3–5.

Classification System uses James Reason's notion of latent and active failures in systems to analyze how mishaps can occur. One of the most significant top-level influences is organizational climate, or culture.[2] If the climate of an organization prevents information flow among different levels, then it is not a matter of *if* trouble will occur, but *when*.[3]

Significantly, the case studies in this volume describe mishaps involving highly qualified people. These accidents were not the result of carelessness or inexperience. If such things can happen to the best and brightest, they can happen to anyone.

---

2. Wiegmann and Shappell, *A Human Error Approach to Aviation Accident Analysis*.

3. Reason, *Human Error*.

Bibliography

Books

Mark R. Anderson and Anthony B. Page, *Multivariable Analysis of Pilot-in-the-Loop Oscillations*, AIAA-95-3203-CP (Reston, VA: AIAA, 1995).

Anon., *Aerospace Walk of Honor* (Lancaster, CA: Davis Communications, 2009).

David C. Aronstein, Michael J. Hirschberg, and Albert C. Piccirillo, *Advanced Tactical Fighter to F-22 Raptor: Origins of the 21st Century Air Dominance Fighter* (Reston, VA: AIAA, 1998).

Richard C. Atkinson and Richard C. Shiffrin, "Human Memory: A Proposed System and Its Control Processes," in *The Psychology of Learning and Motivation: Advances in Research and Theory*, vol. 2, ed. K. Spence and J. Spence (New York: Academic Press, 1968).

Robert D. Banks, James W. Brinkley, Richard Allnutt, and Richard M. Harding, "Human Response to Acceleration," chapter 4 in *Fundamentals of Aerospace Medicine*, ed. Jeffrey R. Davis, Robert Johnson, Jan Stepanek, and Jennifer A. Fogarty, 4th ed. (Philadelphia, PA: Lippincott Williams & Wilkins, 2008), pp. 83–109.

Michael R. Barratt and Sam L. Pool, *Principles of Clinical Medicine for Space Flight* (New York: Springer, 2008).

Jay C. Buckey, *Space Physiology* (New York: Oxford University Press, 2006).

Brian Burrough, *Dragonfly: NASA and the Crisis Aboard Mir* (New York: Harper-Collins, 1998).

Mary Conners, Albert Harrison, and Faren Akins, *Living Aloft—Human Requirements for Long-Duration Spaceflight*, NASA SP–483 (Washington, DC: NASA, 1985).

Henry S.F. Cooper, Jr., *A House in Space* (New York: Holt, 1976).

Steven Cushing, *Fatal Words: Communication Clashes and Aircraft Crashes* (Chicago, IL: University of Chicago Press, 1997).

Roy L. DeHart and Jeffrey R. Davis, eds., *Fundamentals of Aerospace Medicine*, third ed. (Philadelphia, PA: Lippincott Williams & Wilkins, 2002).

Holger Duda, *Effects of Rate Limiting Elements in Flight Control Systems—A New PIO Criterion*, AIAA-95-3204-CP (Reston, VA: AIAA, 1995).

Michael H. Gorn, *Expanding the Envelope: Flight Research at NACA and NASA* (Lexington, KY: The University Press of Kentucky, 2001).

Rex D. Hall and David J. Shayler, *Soyuz: A Universal Spacecraft* (Chichester, U.K.: Praxis Publishing Ltd., 2003).

Richard P. Hallion, *Test Pilots: The Frontiersmen of Flight* (Garden City, NY: Doubleday, 1981).

Richard P. Hallion and Michael H. Gorn, *On the Frontier: Experimental Flight at NASA Dryden* (Smithsonian Books, 2003).

Albert A. Harrison, *Spacefaring: The Human Decision* (Berkeley, CA: University of California Press, 2001).

Albert A. Harrison, Y.A. Clearwater, and Christopher P. McKay, eds., *From Antarctica to Outer Space* (New York: Springer, 1991).

David K. Hitt, Owen K. Garriott, and Joseph P. Kerwin, *Homesteading Space— The Skylab Story* (Lincoln, NE: University of Nebraska Press, 2008).

Irving L. Janis, *Groupthink: Psychological Studies of Policy Decisions and Fiascos* (Chicago, IL: Houghton and Mifflin, 1982).

Dennis R. Jenkins, *X-15: Extending the Frontiers of Flight*, SP-2007-562 (Washington, DC: NASA, 2007).

Dennis R. Jenkins and Tony R. Landis, *Hypersonic: The Story of the North American X-15* (North Branch, MN: Specialty Press, 2002).

Dennis R. Jenkins and Tony R. Landis, *Valkyrie: North American's Mach 3 Superbomber* (North Branch, MN: Specialty Press, 2004).

Nick Kanas and Dietrich Manzey, *Space Psychology and Psychiatry* (El Segundo, CA: Springer and Microcosm Press, 2008).

Frank J. Landy, *Psychology of Work Behavior* (Pacific Grove, CA: Brooks/Cole, 1989).

Wiley J. Larson and Linda K. Pranke, eds., *Human Spaceflight* (New York: McGraw-Hill, 2007).

Jerry M. Linenger, *Off the Planet: Surviving Five Perilous Months Aboard the Space Station Mir* (New York: McGraw-Hill, 2000).

Don Logan, *Rockwell B-1B: SAC's Last Bomber* (Atglen, PA: Schiffer Military History, 1995).

Maura Phillips Mackowski, *Testing the Limits: Aviation Medicine and the Origins of Manned Space Flight* (College Station, TX: Texas A&M University Press, 2006).

Donald L. Mallick and Peter W. Merlin, *The Smell of Kerosene: A Test Pilot's Odyssey,* NASA SP-4108 (Washington, DC: NASA, 2003).

Mike D. McNeese, Ed Salas, and Mica Endsley, eds., *New Trends in Cooperative Activities* (Santa Monica, CA: Human Factors and Ergonomics Society, 2001).

Peter W. Merlin and Tony Moore, *X-Plane Crashes—Exploring Secret, Experimental, and Rocket Plane Crash Sites* (North Branch, MN: Specialty Press, 2008).

Jay Miller, *The X-Planes: X-1 to X-45* (Hinckley, U.K.: Midland Publishing, 2001).

Clay Morgan, *Shuttle-Mir: The United States and Russia Share History's Highest Stage,* NASA SP-4225 (Washington, DC: NASA, 2001).

James D. Murphy, *Business Is Combat* (New York: Harper Collins, 2000).

Bill Ocker and Carl Crane, *Blind Flight Guidance* (San Antonio, TX: Naylor Book Publishing, 1932).

Steve Pace, *Boeing North American B-1 Lancer* (North Branch, MN: Specialty Press, 1998).

Donald E. Parker and K.L. Parker, "Adaptation to the Simulated Stimulus Arrangement of Weightlessness," in *Motion and Space Sickness*, ed. G. Crampton (Boca Raton, FL: CRC Press, 1990), pp. 247–262.

William R. Pogue, *How Do You Go to the Bathroom in Space?* (New York: Tom Doherty Associates, 1985).

James Reason, *Human Error* (New York: Cambridge University Press, 1990).

James Reason, *Managing the Risks of Organizational Accidents* (Aldershot, U.K.: Ashgate Publishing Ltd., 1997).

R. Dale Reed and Darlene Lister, *Wingless Flight: The Lifting Body Story*, NASA SP-4220 (Washington, DC: NASA, 1997).

Jeannette Remak and Joe Ventolo, Jr., *XB-70 Valkyrie: The Ride to Valhalla* (St. Paul, MN: MBI Publishing Company, 1998).

Linda Shackelford, "Musculoskeletal Response to Space Flight," in *Principles of Clinical Medicine for Spaceflight*, ed. Barratt and Pool (New York: Springer, 2008), pp. 293–306.

David J. Shayler, *Disasters and Accidents in Manned Spaceflight* (Chichester, U.K.: Praxis Publishing Ltd., 2000).

Thomas B. Sheridan, *Telerobotics, Automation, and Human Supervisory Control* (MIT: Cambridge Press, 1992).

Stedman's Medical Dictionary, 27th ed. (Philadelphia, PA: Lippincott, Williams & Wilkins, 2000).

Jack Stuster, *Bold Endeavors: Lessons from Polar and Space Exploration* (Annapolis, MD: Naval Institute Press, 1996).

Loyd S. Swenson, Jr., James M. Grimwood, and Charles C. Alexander, *This New Ocean: A History of Project Mercury*, SP-4201 (Washington, DC: NASA, 1966; repr. 1999).

Donald I. Tepas, M.J. Paley, and S.P. Popkin, "Work Schedules and Sustained Performance," in *Handbook of Human Factors*, ed. G. Salvendy (New York: Wiley, 1997) pp. 1021–1058.

Milton O. Thompson, *At the Edge of Space: The X-15 Flight Program* (Washington, DC: Smithsonian Institution Press, 1992).

Milton O. Thompson and Curtis Peebles, *Flying Without Wings: NASA Lifting Bodies and the Birth of the Space Shuttle* (Washington, DC: Smithsonian Institution Press, 1999).

Christopher D. Wickens, John D. Lee, Yili Liu, and Sallie E. Gordon Becker, *An Introduction to Human Factors Engineering*, second ed. (Upper Saddle River, NJ: Pearson Prentice Hall, 2004).

Douglas A. Wiegmann and Scott A. Shappell, *A Human Error Approach to Aviation Accident Analysis: The Human Factors Analysis and Classification System* (Aldershot, U.K.: Ashgate Publishing, 2003).

Jon Weimer, ed., *Research Techniques in Human Engineering* (New York: Prentice Hall, 1995).

Earl L. Weiner and David C. Nagel, *Human Factors in Aviation* (San Diego, CA: Academic Press, 1988).

John A. Wise, V. David Hopkin, and Daniel J. Garland, *Handbook of Aviation Human Factors* (Boca Raton, FL: CRC Press, 2010).

Tom Wolfe, *The Right Stuff* (New York: Farrar, Straus and Giroux, 1979).

Journal Articles and Technical Papers

Albert J. Ahumada, Maite Trujillo San-Martin, and Jennifer Gille, "Symbol Discriminability Models for Improved Flight Displays," *SPIE Proceedings*, vol. 6057, paper 30 (January 2006).

Jeffrey J. Armentrout, Dwight A. Holland, Kevin J. O'Toole, and William R. Ercoline, "Fatigue and Related Human Factors in the Near Crash of a Large Military Aircraft," *Aviation, Space, and Environmental Medicine*, vol. 77, no. 9 (2006), pp. 963–970.

Rashid L. Bashshur and Corinna E. Lathan, "Human Factors in Telemedicine," *Telemedicine Journal*, vol. 5, no. 2 (July 1999), pp. 127–128.

B.J. Bluth, "The Benefits and Dilemmas of an International Space Station," *Acta Astronautica*, vol. 11 (1984), pp. 149–153.

B.J. Bluth, "Soviet Space Stress," *Science*, no. 2 (1982), pp. 30–35.

Adam R. Brody, "Spacecraft Flight Simulation: A Human Factors Investigation into the Man-Machine Interface Between an Astronaut and a Spacecraft Performing Docking Maneuvers and Other Proximity Operations," master's thesis, MIT, Cambridge, MA (April 1987); see also NASA CR-177502 (September 1988).

Adam R. Brody, "Evaluation of the '0.1% Rule' for Docking Maneuvers," *Journal of Spacecraft and Rockets*, vol. 27 (1990), pp. 7–8.

Adam R. Brody, R.H. Jacoby, and Stephen R. Ellis, "Extravehicular Activity Self Rescue Using a Hand Held Thruster," *Journal of Spacecraft and Rockets*, vol. 29, no. 6 (1992), pp. 842–848.

J. Christensen and J. Talbot, "Psychological Aspects of Space Flight," *Aviation, Space, and Environmental Medicine*, vol. 57 (1986), pp. 203–212.

Daniel A. Cohen, Alvaro Pascual-Leme, Daniel Z. Press, and Edwin M. Robertson, "Off-Line Learning of Motor Skill Memory: A Double Dissociation of Goal and Movement," Proceedings of the National Academy of Sciences, no. 102 (2005), pp. 18237–18241.

P. De Weerd, K. Reinke, L. Ryan, T. McIsaac, P. Perschler, D. Schnyer, T. Trouard, and A. Gmitro, "Cortical Mechanisms for Acquisition and Performance of Bimanual Motor Sequences," *Neuroimage*, no. 4 (2003), pp. 1405–1416.

Stephen R. Ellis, "Collision in Space: Human Factors Elements of the Mir-Progress 234 Collision," *Ergonomics in Design*, vol. 8, no. 1 (January 2000).

M.R. Endsley, "Toward a Theory of Situation Awareness in Dynamic Systems," *Human Factors,* vol. 37 (1995).

S.J. Gerathewohl and H.D. Stallings, "Experiments During Weightlessness: A Study of the Oculoagravic Illusion," *Journal of Aviation Medicine*, no. 29 (1958).

Ashton Graybiel, "Oculogravic Illusion," *Archives of Ophthalmology*, vol. 48, no. 5 (1952).

Arthur J. Grunwald and Stephen R. Ellis, "A Visual Display Aid for Orbital Maneuvering: Design Considerations," *AIAA Journal of Guidance and Control*, vol. 16, no. 1 (1993), pp. 139–144.

Mark A. Guidi and Dwight A. Holland, "Operational and Human Factors Implications of Physiological Deconditioning in Long Duration Spaceflight," *Aerospace Systems*, Human Factors and Ergonomics Society Annual Meeting Proceedings (1992), pp. 116–120.

A.A. Harrison and M.M. Conners, "Groups in Exotic Environments," *Advances in Experimental Social Psychology*, no. 18 (1984), pp. 49–87.

Robert L. Helmreich, "Culture and Error in Space: Implications from Analog Environments," *Aviation, Space, and Environmental Medicine*, vol. 71, no. 9, section 2, suppl. (September 2000), pp. A133–A139.

Robert L. Helmreich, A.C. Merritt, and J.A. Wilhelm, "The Evolution of Crew Resource Management Training," *International Journal of Aviation Psychology*, vol. 9, no. 1 (1999), pp. 19–32.

Dwight A. Holland, "A Systems Case Study Examination of the Near-Catastrophic Mir-Progress 234 Collision with Emphasis on the Human Factors Issues Surrounding This Mishap," Proceedings of the INCOSE Mid-Atlantic Regional Conference, Reston, VA, April 5–8, 2000, pp. 18.2-1–18.2-8.

Dwight A. Holland and William R. Barfield, "Some Virtual Reality and Telemedicine Applications Useful for Long Duration Spaceflight from a Systems Engineering Perspective," ed. Westwood, Hoffman, Robb, and Stredney, *"Medicine Meets Virtual Reality"* (Washington, DC: IOS Press, 1999), pp. 141–147.

D.H. Hull, R.A. Wolthuis, K.K. Gillingham, and J.H. Triebwasser, "Relaxed +G_z Tolerance in Healthy Men: Effect of Age," *Journal of Applied Physiology*, vol. 45, no. 4 (1978), pp. 626–629.

Tovey Kamine and Gregg A. Bendrick, "Visual Angles of Conventional and a Remotely Piloted Aircraft," *Aviation, Space, and Environmental Medicine* (2009).

Nick Kanas, "Psychological and Interpersonal Issues in Space," *American Journal of Psychiatry*, vol. 144 (1987), pp. 703–709.

Sophie Lalande and Fred Buick, "Physiologic +G_z Tolerance Responses over Successive +G_z Exposures in Simulated Air Combat Maneuvers," *Aviation, Space, and Environmental Medicine*, vol. 80, no. 12 (December 2009).

Brad S. Liebst et al., "Nonlinear Pre-filter To Prevent Pilot-Induced Oscillations Due to Actuator Rate Limiting," *AIAA Journal of Guidance, Control, and Dynamics*, vol. 25, no. 4 (2002), pp. 740–747.

T.J. Lyons, W.R. Ercoline, J.E. Freeman, and K.K. Gillingham, "Classification Problems of U.S. Air Force Spatial Disorientation Accidents, 1989–91," *Aviation, Space, and Environmental Medicine*, vol. 65 (1994).

S.P. McKee, "A Local Mechanism for Differential Velocity Detection," *Vision Research*, vol. 21 (1981), pp. 491–500.

Peter W. Merlin, "Free Enterprise: Contributions of the Approach and Landing Test (ALT) Program to the Development of the Space Shuttle Orbiter," presented at AIAA Space Conference, San Jose, CA, September 21, 2006, AIAA-2006-7467.

G. Miller, "The Magical Number Seven, Plus or Minus Two: Some Limits on Our Capacity for Processing Information," *Psychological Review*, vol. 63, no. 2 (1956), pp. 81–97.

Thomas W. Moore, Dov Jaron, Leonid Hrebien, and David Bender, "A Mathematical Model of G Time-Tolerance," *Aviation, Space, and Environmental Medicine*, vol. 64 (1993), pp. 947–951.

K.L. Morrissette and D.G. McGowan, "Further Support for the Concept of a G-LOC Syndrome: A Survey of Military High-Performance Aviators," *Aviation, Space, and Environmental Medicine*, vol. 71 (2000), pp. 496–500.

John M. Nicholas, "Interpersonal Issues and Group Behavior Skills Training for Crews on Space Station," *Aviation, Space, and Environmental Medicine*, vol. 60 (1989), pp. 603–608.

John M. Nicholas, H. Clayton Foushee, and Francis L. Ulschak, *Crew Productivity Issues in Long Duration Spaceflight*, report AIAA-88-04444 (Washington, DC: AIAA, 1988).

John M. Nicholas and L. Penwell, "A Proposed Profile of the Effective Leader in Human Spaceflight Based upon Findings from Analog Environments," *Aviation, Space, and Environmental Medicine*, vol. 66 (1995), pp. 63–72.

James E. Oberg, "Shuttle-Mir's Lessons for the International Space Station," *IEEE Spectrum* (June 1998), pp. 28–37.

Bruce G. Powers, "Low-Speed Longitudinal Orbiter Flying Qualities," Space Shuttle Technical Conference, pt. 1, NASA CP-2342, Houston, TX, Johnson Space Center, 1983.

Fred H. Previc and William R. Ercoline, eds., "Spatial Disorientation in Aviation," *Progress in Astronautics and Aeronautics*, vol. 203 (Reston, VA: AIAA, 2004).

Caroline A. Rickards and David G. Newman, "G-Induced Visual and Cognitive Disturbances in a Survey of 65 Operational Fighter Pilots," *Aviation, Space, and Environmental Medicine*, vol. 76, no. 5 (2005), pp. 496–500.

Barry A. Romich, "Knowledge in the World Vs. Knowledge in the Head: The Psychology of AAC Systems," *Communication Outlook*, vol. 16, no. 2 (Pittsburgh, PA: AAC Institute Press, 1994).

Michael B. Russo, Helen C. Sing, Saul Santiago, Athena P. Kendall, Dagny E. Johnson, David R. Thorne, Sandra M. Escolas, Dwight A. Holland, Stanley W. Hall, and Daniel P. Redmond, "Occurrence and Patterns of Visual Neglect in U.S. Air Force Pilots in a Simulated Overnight Flight," *Aviation, Space, and Environmental Medicine*, vol. 75, no. 4, section 1 (2004), pp. 323–332.

Michael B. Russo, Athena P. Kendall, Dagny E. Johnson, Helen C. Sing, Sandra M. Escolas, Saul Santiago, Dwight A. Holland, Stanley W. Hall, and Daniel P. Redmond, "Visual Perception, Psychomotor Performance, and Complex Motor Performance During an Overnight Air Refueling

Simulated Flight," *Aviation, Space, and Environmental Medicine*, vol. 76, no. 7, section II (2005), pp. 92–103.

Eduardo Salas, C. Prince, David P. Baker, and L.B. Shrestha, "Situation Awareness in Team Performance: Implications for Measurement and Training," *Human Factors*, vol. 37 (Santa Monica, CA: 1995), pp. 123–136.

P. Santy, "The Journey In and Out: Psychiatry and Space Exploration," *American Journal of Psychiatry*, vol. 140 (1983), pp. 519–527.

Wg. Cdr. A. Sinha and Wg. Cdr. P.K. Tyagi, "Almost Loss of Consciousness (A-LOC): A Closer Look at Its Threat in Fighter Flying," *Indian Journal of Aerospace Medicine*, vol. 48, no. 2 (2004), pp. 17–21.

Barry S. Shender, Estrella M. Forster, Leonid Hrebien, Han Chool Ryoo, and Joseph P. Cammarota, Jr., "Acceleration-Induced Near-Loss of Consciousness: The 'A-LOC' Syndrome," *Aviation, Space, and Environmental Medicine*, vol. 74, no. 10 (2003), pp. 1021–1028.

Alice M. Stoll, "Human Tolerance of Positive G as Determined by the Physiological End Points," *Journal of Aviation Medicine*, vol. 27 (1956), pp. 356–367.

L.S. Stone and P. Thompson, "Human Speed Perception Is Contrast Dependent," *Vision Research*, vol. 32 (1992), pp. 1535–1549.

Anthony P. Tvaryanas, William T. Thompson, and Stefan H. Constable, "Human Factors in Remotely Piloted Aircraft Operations: HFACS (Human Factors Analysis Classification System) Analysis of 221 Mishaps over 10 Years," *Aviation, Space, and Environmental Medicine*, no. 77 (2006), pp. 724–732.

H. Van Dongen, "Comparison of Mathematical Model Predictions to Experimental Data of Fatigue and Performance," ed. Neri and Nunneley, Proceedings of the Fatigue and Performance Modeling Workshop, Seattle, WA, June 13–14, 2002, *Aviation, Space, and Environmental Medicine*, vol. 75, no. 3, section II (2004), pp. A15–36.

P.M. van Wulfften Palthe, "Function of the Deeper Sensibility and of the Vestibular Organs in Flying," *Acta Otolaryngologica*, vol. 4 (1922).

Earl Wiener, "Controlled Flight into Terrain Accidents: System-Induced Errors," *Human Factors*, vol. 19 (1997), pp. 171–181.

D.B. Willingham, M.J. Nissen, and P. Bullemer, "On the Development of Procedural Knowledge," *Journal of Experimental Psychology, Learning, Memory, and Cognition*, no. 6 (1989), pp. 1047–1060.

Periodicals

Brian Burrough, "Letter from Space: All Heaven in a Rage," *Vanity Fair* (November 1988), pp. 132 ff.

Andrew Chaikin, "The Loneliness of the Long-Distance Astronaut," *Discover* (February 1985), pp. 20–31.

Michael A. Dornheim, "X-31 Board Cites Safety Analyses, but Not All Agree," *Aviation Week & Space Technology* (December 4, 1995).

Jeffrey Kluger, "How I Survived Mir: A Bad Day in Space," *Time* (November 3, 1997), pp. 84–91.

Peter W. Merlin, "Michael Adams: Remembering a Fallen Hero," *The X-Press*, vol. 46, no. 6 (July 30, 2004).

Peter W. Merlin, "The Real Six Million Dollar Man—Bruce Peterson," *World X-Planes*, no. 3 (2006).

Victor Riley, "Reducing Mode Errors Through Design," *Avionics Magazine* (March 1, 2005).

Jon Thurber, "David P. Cooley Dies at 49; Test Pilot Worked for Air Force, Lockheed Martin Before Fatal Crash," *Los Angeles Times* (March 30, 2009).

Internet Sites

"F-22 Raptor Cockpit," GlobalSecurity.org, *http://www.globalsecurity.org/military/systems/aircraft/f-22-cockpit.htm*, accessed March 20, 2010.

SrA. Jason Hernandez, "Raptor Drops First Small Diameter Bomb," F-16. net, *http://www.f-16.net/news_article2528.html*, accessed March 20, 2010.

Jeff Hollenbeck, "F-22 Raptor Team Delivers the Last Developmental Flight-Test Aircraft to USAF," F-16.net, *http://www.f-16.net/news_article1670. html*, June 3, 2003, accessed March 20, 2010.

Human Factors and Ergonomics Society, *http://www.hfes.org/web/AboutHFES/ about.html*, accessed April 6, 2010.

Mars Society Web site, *http://www.mars.org*, accessed July 10, 2007.

NASA, "Lessons Learned for Phase I International Space Station," NASA Human Space Flight, *http://spaceflight.nasa.gov/history/shuttle-mir/history/ h-b-lessons.htm*, 1988, accessed July 10, 2007.

Peter W. Merlin, "Research Data from the X-15 Program Contributed to Apollo Lunar Missions," NASA Dryden Flight Research Center News Features, *http://www.nasa.gov/centers/dryden/Features/X-15X-15_Apollo.html*, uploaded July 10, 2009, accessed June 23, 2010.

Public Broadcasting System, "Terror in Space," NOVA transcript no. 2513, PBS, WGBH Boston, *http://www.pbs.org/wgbh/nova/mir*, 1998.

SrA. Julius Delos Reyes, "F-22 Raptor Performs First Supersonic SDB Drop," F-16.net, *http://www.f-16.net/news_article2975.html*, accessed March 20, 2010.

Asif Shamim, "F-22 Flameout During SDB Flight Testing," F-16.net, *http://www.f-16.net/news_article2539.html*, October 1, 2007, accessed March 20, 2010.

C. van den Berg, "Mir News," InfoThuis.nl, *http://infothuis.nl/muurkrant/ mirmain.html*, 1999, accessed August 3, 2010.

James E. Whinney, Ph.D., M.D., Aeromedical Research Division, Civil Aerospace Medical Institute, "Sustained Acceleration Exposure," *Advanced Aerospace Medicine On-line*, section II.2.7, *http://www.faa.gov/other_visit/ aviation_industry/designees_delegations/designee_types/ame/tutorial/*, updated September 29, 2005, accessed February 18, 2010.

Government Reports

William H. Andrews, "Space Shuttle Orbiter Approach & Landing Test—Mated Inert Flight Test Plan," NASA DFRC, January 28, 1977.

Approach and Landing Test Evaluation Team, "Space Shuttle Orbiter Approach and Landing Test—Final Evaluation Report," JSC-13864, February 1978.

Donald R. Bellman et al., "Investigation of Landing Accident with the M2-F2 Lifting Body Vehicle on May 10, 1967, at Edwards, California," NASA Flight Research Center, June 1967.

Donald R. Bellman et al., "Investigation of the Crash of the X-15-3 Aircraft on November 15, 1967," NASA Flight Research Center, 1968.

Marvin E. Bonner, Edward Browning, Arthur Cobb, Travis Masters, Gary Middleton, Robert D. Murphy, Don M. Springman, and John Van Schaik, *Tactical Aircraft: F-22 Delays Indicate Initial Production Rates Should Be Lower To Reduce Risks,* GAO-02-298, March 2002.

C.R. Chalk and P.A. Reynolds, "Test Plan—Simulation of Orbiter Landing Characteristics in the USAF Total In-Flight Simulator (TIFS)," Calspan Corporation, TIFS Memo no. 844, May 25, 1978.

Department of Defense, *Flying Qualities of Piloted Aircraft*, MIL-HDBK-1797, 1997.

Maj. Gen. David W. Eidsaune, *United States Air Force Accident Investigation Board Report: F-22A, T/N 91-4008*, U.S. Air Force, July 2009.

FAA, Office of the Chief Scientific and Technical Advisor for Human Factors, *Guidelines for Human Factors Requirements Development*, AAR-100, version 1.0, February 6, 2003.

FAA, FAA Advisory Circular AC No. 120-51E, "Crew Resource Management Training," January 22, 2004.

Robert Hoey, "Testing Lifting Bodies at Edwards," in *Air Force/NASA Lifting Body Legacy History Project* (Lancaster, CA: PAT Projects, Inc., 1994).

Robert G. Hoey et al., "Flight Test Results from the Entry and Landing of the Space Shuttle Orbiter for the First Twelve Orbital Flights," AFFTC-TR-85-11, June 1985.

Investigation of USAF aircraft accident, August 29, 1984, B-1A 74-0159, 1984.

Robert W. Kempel, "Analysis of a Coupled-Roll-Spiral Mode, Pilot Induced Oscillation Experienced with the M2-F2 Lifting Body," NASA TN D-6496, 1971.

David H. Klyde et al., "Unified Pilot-Induced Oscillation Theory," vol. 1, "PIO Analysis with Linear and Nonlinear Effective Vehicle Characteristics, Including Rate Limiting," WL-TR-96-3028 (Air Force Research Laboratory, Wright-Patterson AFB, OH, 1995).

Larry P. Krock, "Aeromedical Issues Related to Positive Pressure Breathing for $+G_z$ Protection," TP-1992-0024, 1993.

Alexei Leonov and Vladimir Lebedev, Psychological Problems of Planetary Flight, NASA Technical Translation F-16536, 1975.

Lyndon B. Johnson Space Center, *Lessons Learned on Skylab Program*, NASA TM X-72920, 1974.

Duane T. McRuer, Seventh Report of the NASA Advisory Council Task Force on the Shuttle-Mir Rendezvous, *Pilot-Induced Oscillations and Human Dynamic Behavior*, NASA Contractor Report CR-4683 (Hawthorne, CA: Systems Technology, Inc., 1995).

NASA 584 X-31 Mishap Investigation Report, NASA Dryden Flight Research Center, DFRC Historical Reference Collection, Edwards, CA, August 18, 1995.

David S.F. Portree, "Mir Hardware Heritage," NASA RP 1357, 1995.

Bruce G. Powers, "Space Shuttle Pilot-Induced-Oscillation Research Testing," NASA TM 86034, February 1984.

NASA, "Press Kit: Space Shuttle Orbiter Test Flight Series," release no. 77-16, 1977.

S.A. Shappell and D.A. Wiegmann, *The Human Factors Analysis and Classification System (HFACS)*, report no. DOT/FAA/AM-00/7, 2000.

John W. Smith, "Analysis of a Longitudinal Pilot-Induced Oscillation Experienced on the Approach and Landing Test of the Space Shuttle," NASA TM-81366, December 1981.

USAF, *Executive Summary: Aircraft Accident Investigation, F-22A, S/N 91-4008, Dobbins ARB, Georgia, 22 April 2002*, 2003.

D. Wagner, J.A. Birt, M. Snyder, and J.P. Duncanson, *Human Factors Design Guide for Acquisition of Commercial-Off-The-Shelf Subsystems, Non-Developmental Items, and Developmental Systems*, DOT/FAA/CT-96/1, January 1996.

Joel B. Witte, "An Investigation Relating Longitudinal Pilot-Induced Oscillation Tendency Rating to Describing Function Predictions for Rate-Limited Actuators," AFIT/GAE/ENY/04-M16, March 2004.

XB-70/F-104 Accident Investigation Report, 1966, NASA Dryden Historical Reference Collection.

Interviews, Correspondence, and Miscellaneous

David P. Cooley, "Brief Summary of Flight Test Career," biographical information supplied to the Society of Experimental Test Pilots, May 2008.

Frank L. Culbertson, Jr., testimony before the Committee on Science, U.S. House of Representatives, Washington, DC, September 18, 1997.

Robert Cummings, personal interview by Gregg Bendrick at NASA Dryden Flight Research Center, March 17, 2004.

"DFRC Orbiter Landing Investigation Team Final Presentation," August 1978, file folder L1-5-7-22, Milton O. Thompson Collection, NASA DFRC Historical Reference Collection.

Charles Gordon Fullerton biographical files, Dryden Historical Reference Collection.

Fred W. Haise, Jr., biographical files, Dryden Historical Reference Collection.

Dwight A. Holland, "Systems and Human Factors Concerns During Long-Duration Spaceflight," M.S. thesis, Virginia Polytechnic Institute and State University, Blacksburg, VA (1991).

Public Broadcasting System, *Terror in Space*, NOVA transcript no. 2513, Boston, MA, 1998.

Herman A. Rediess, "Assessment of ALT Tests with Tailcone On Vs. Off," memorandum to DFRC Shuttle Projects Manager from Director of Research, March 17, 1976, file folder L2-1-3C-6, Space Shuttle files, NASA DFRC Historical Reference Collection.

X-31 ITO memo, September 6, 1994, X-31 correspondence files-personnel, DFRC Historical Reference Collection.

Nomination and citation documents, NASA Public Service Medal for Karl-Heinz Lang, X-31 correspondence files-personnel, DFRC Historical Reference Collection.

Lt. Gen. Richard V. Reynolds (USAF, retired), interview at NASA Dryden Flight Research Center, Edwards, CA, July 15, 2008.

M. Ridgeway, "Leadership," a section in the 2002 Course 27 Squadron Officer School course materials, October 1966; reprinted in 2002 with permission from *Military Review,* Maxwell AFB, AL, pp. 40–49.

Lt. Gen. Thomas P. Stafford, testimony before the Committee on Science, House of Representatives, Washington, DC, May 6, 1998.

Milton O. Thompson, notes on "Sequence of Events on FF-5 Landing," n.d., file folder L1-5-2-2, Milton O. Thompson Collection, NASA DFRC Historical Reference Collection.

Col. Art Tomassetti (USMC), "Flight Test 2040," presented at the 54th Annual Society of Experimental Test Pilots Symposium, Anaheim, CA, September 25, 2010.

Index

Numbers in **bold** indicate pages with illustrations.

747 airliner
  as Shuttle Carrier Aircraft (SCA), **37**, 38–40,
    44
  Tenerife crash, 19

A

Abrams, Richard, 145
acceleration
  control order of systems, 34
  g-forces, 101–2
  input and, 34
  oculoagravic illusion and, 55, 67–68, 69,
    73, 77, 78
  oculogravic illusion and, 67
  pitch-up illusion and, 66–67
  somatogravic illusion and, 66–67
  Space Shuttle cues of, 46, 48
  unit of acceleration forces (g), 112
ACES II ejection seat/system, 103, 110
active errors/failures, xiii, **xiv**, 190
Adams, Michael J.
  education, training, and experience of,
    59–62, **60**
  F-104 landing accident, 60–61
  medical screening tests, 68, 69, 77
  oculoagravic illusion, 55, 67–68, 69, 73,
    77, 78
  vertigo (spatial disorientation), 62, 65–70, 77
  X-15 accident, 62–65, **65**, 68–69, 70–73,
    72
  X-15 flights, 60, 61–62
Advanced Manned Strategic Aircraft study, 143
Advanced Tactical Anti-G Suit (ATAGS), 120

Advanced Tactical Fighter (ATF) program, 102.
  See also F-22A Raptor
age and G-LOC, 122
AGM-86B cruise missile, 146
aileron-rudder interconnection linkage, 26, 27,
  28–30, 86, 91–92
AIM-9X missile, **108**
AIM-120D air-to-air missile, 104
air combat maneuvering
  g-forces and, 102, 121
  situational awareness and, 102
aircraft
  advances in, x
  design of, focus of, ix
  flight performance, functions that impact, ix
  maneuverability of, 3, 5
  See also flight research and testing
aircraft-pilot coupling, 35. See also pilot-induced
  oscillation (PIO)
Air Force, U.S.
  A-LOC situations, 117
  age and g-tolerance, study of, 122
  B-1 Combined Test Force (CTF), 145, 146,
    147, 149
  GE engine publicity photos, 132–41, **134**,
    136, **137**, **138**
  Man-in-Space Soonest project, 130–31
  Manned Orbiting Laboratory (MOL), 42, 61,
    77
  mishaps and human factors, 189
  strategic bomber force, 143
  Test Pilot School, 84, 107
  Total In-Flight Simulator (TIFS), 47–49

31 X-15 flight research program, 55

X-31 program, 5

XB-70A Valkyrie program, 128, 129. *See also* B-70 Valkyrie and XB-70A Valkyrie bomber prototype

Air Force Flight Test Center, Bioastronautics Branch, 59

airframe/powerplant configuration, flight performance and, ix

Allen, H. Julian, 24

almost-loss of consciousness (A-LOC), 102, 111, 117–18, 121, 123, 189

altitude

    record flights, 56, 57, 129, 131

    Space Shuttle cockpit visibility and judgment of, 48

    X-15 high-altitude flights, 56, 57

Alvarez, Vincent A., 38

Ames Aeronautical Laboratory, 24

angles of attack (AOA)

    Kiel probe and air data, 7, 8, 9, 13, 16, 19, 20, 21

    X-31 and high AOA, 3–6, **4**

anti-g straining maneuver (AGSM), 111, 118–19, 120–21, 122

Apollo program

    ballistic entry and crew positioning, 23

    contractor for Apollo Command Service Module, 58

    Fullerton's missions, 42–43

    Haise's missions, 42

    Lunar Landing Research Vehicle, 131

    reentry, aerodynamic heating, and thermal protection, 58

    Saturn V rocket, 58

    Scott's role in, 61

    windshield design, 58

    X-15 flight data and Apollo spacecraft design, 58

Approach and Landing Test (ALT) Program

    captive flights, 38, 39

    crews and pilots for test flights, 38, **39**

    Enterprise orbiter vehicle prototype, **32**, 37–38, **37**

    free flight and landing PIO incident, 43–46, **43**

    free flights, 38, 39–41, **40**

    PIO condition, operational fix for, 47–49

    purpose of, 36, 38

    Shuttle Carrier Aircraft (SCA) launch, **37**, 38–40, 44

    taxi tests, 38–39

    test site, 36–37

Armstrong, Neil, 56

Army Air Corps pilot screening tests, 68, 76–77

asthenia, 171

astronauts

    astronaut wings, flights to earn, 56–57, 62, 65, 131

    bone-density losses in space, 169

    exercise protocols, 169–70, 174, 185

    medical screening tests, 68–69, 76–77

    muscle-mass losses in space, 169, 170–71

ATAGS (Advanced Tactical Anti-G Suit), 120

Atlantis, 165

attention

    channelized attention (cognitive tunneling), 95, 97

    distraction and, 94–98

    divided attention, 96–97

    focused (sustained) attention, 96

    radar scanner hypothesis, 97

    selective attention, 96, 98

    situational awareness and, 98

    workload and, 98–99

Attention-Based Principles, 74

attitude indicator, 55, 63, 70–73, **70**, **72**, 75

automation bias

    concept of, 13

    X-31 accident and, 3, 10–13

B

B-1 bomber
  Air Vehicle One (AV-1), 145
  Air Vehicle Two (AV-2), 145–46, 149
  Air Vehicle Four (AV-4), 145
  B-1A aircraft, 145, **154**
  B-1B aircraft, 145–46
  cancellation of production, 145
  caution panel and warning fatigue, 152, **154**, 156–57
  communications, 157
  construction and materials, 143–44
  cost of, 145, 155
  crew resource management, 155, **156**
  design and characteristics of, **142**, 143–44, **144**, **151**
  ejection seats, 145
  emergency systems, 145
  engines, **142**, 144
  escape capsule, 145, **153**, 154–55, 157
  first flight, 145
  Flight 2-127 test flight, 146–47, 148–57, **156**
  flight control system, 146
  fuel transfer system, 144, 153, 155, 156
  low-altitude flights, **142**, 144
  low-speed maneuvers, 143, **144**, 146, 151–52
  number of prototypes and aircraft, 143
  record flight, 145
  reinstatement of production, 145
  safety data and parameters, monitoring of, 149–50
  stability control augmentation system, 153
  test profile tasking (test card), 146
  weapon separation tests, 146, 150
  wing-sweep maneuvers, center of gravity, and stability, 144, 149, 151–53, **151**, 155–56, 157

B-1 Combined Test Force (CTF), 145, 146, 147, 149
B-2 Advanced Technology Bomber, 90–91
B-29 bomber, 130
B-52 Stratofortress
  cockpits, x
  M2-F2 mother ship launch, 26, 27–28, **27**, 84, 85–86
  replacement of, 143
  X-15 mother ship launch, 27, 56, 63
B-58 aircraft, 133, 136, 139, 141
B-70 Valkyrie
  cancellation of, 128
  design and characteristics of, 127
  focus of program, 128
  funding for program, 127–28
  See also XB-70A Valkyrie bomber prototype
B61 nuclear weapon shape, 146
B83 nuclear weapon shape, 146
Bartlett, Frederick, x
Bates, Richard L., 148–49
Bell X-1/X-1A aircraft, **viii**, ix, 130
Bellman, Donald R., 65
Benefield, Tommie D. "Doug"
  B-1 bomber Flight 2-127 test flight, 146–47, 148–49, 150–57
  education, training, and experience of, 145, 147–49, **148**, **156**
Bikle, Paul F., 24
biomedical data, 59
blackout, 101, 114–16, **115**
Blaha, John, 170–71, 173, 174
Blind Flight Guidance (Ocker and Crane), x
blunt-body aerodynamics, 24
Bock, Charles C., Jr., 145
Boeing 747 airliner
  as Shuttle Carrier Aircraft (SCA), **37**, 38–40, 44
  Tenerife crash, 19
bomber aircraft

development of long-range, high-
performance, 127
strategic bomber force, 143
See also specific aircraft
bone-density losses in space, 169
Bowline, Jerry D., 27, 85
Brooks AFB, 68, 69, 77
Brooks Field, 76
Broughton, Robert, 149–50
Brown, Harold, 140

C

C-47 aircraft, 84
C-131H aircraft (Calspan Total In-Flight
Simulator), 47–49
C-133 cargo plane, 148
Calspan Total In-Flight Simulator (TIFS), 47–49
Cambridge University, Cambridge Cockpit, x
canards, 3, 128
cardiovascular system and g-forces, 112–13,
115, 116, 121
Carter, Jimmy, 145
Cate, Albert, 132, 140
caution panel and warning fatigue, 152, **154**,
156–57
centrifugal force, 101, 112, 113
centrifuge studies, 23, 115–16, 117, 122
Challenger, 43
channelized attention (cognitive tunneling), 95,
97. *See also* attention
closed-loop instability, 33, 35, 89. *See also* pilot-
induced oscillation (PIO)
cockpits
caution panel and "warning fatigue," 152,
154, 156–57
design of, study and understanding of, xiii
design of human-vehicle interface, 78, 189
glass cockpits, xi, **xii**, 103
size of, 103
visibility from, 46, 48

See also instruments and instrument panels
cognitive function
almost-loss of consciousness (A-LOC), 102,
111, 117–18, 121, 123, 189
asthenia, 171
attention and distraction, 94–98. *See also*
attention
flight tests and challenges to, ix
g-forces and, 102, 189
g-induced loss of consciousness (G-LOC),
101–2, 112, 113, 114–18, **115**, 119,
121, 122, 189
sleep deprivation and, 182–83
visual-information processing, 95–97
cognitive tunneling (channelized attention), 95,
97. *See also* attention
Collier Trophy, 131
Columbia, 48, 50–51, **50**
Combined Advanced Technology Enhancement
Design G Ensemble (COMBAT EDGE) positive-
pressure breathing and counter-pressure jerkin
vest, 119–20, 122
communications
B-1 bomber Flight 2-127 test flight, 157
complacency and, 18–19
control room culture and, 18
F-22A Raptor, 109, 111
fatigue and, 18–19
M2-F2 lifting body, 87
Mir space station, 173, 175–76, 180–81,
184
mishaps and, 189
NASA-RSA communications, 180–81
organizational communications philosophy,
18, 190
study and understanding of, xiii
successful missions, systems and human
factors for, 187, **187**
X-31 accident and, 3, 7–8, 9, 14–15, 16,
18–21

complacency and communications, 18–19
computer technology, xi–xii
Concorde SST, 148
consciousness
    almost-loss of consciousness (A-LOC), 102,
        111, 117–18, 121, 123, 189
    blackout, 101, 114–16, **115**
    g-forces and, 101–2
    g-induced loss of consciousness (G-LOC),
        101–2, 112, 113, 114–18, **115**, 119,
        121, 122, 189
    grayout, 101, 112, 114–16, **115**
control surfaces
    Enterprise orbiter vehicle prototype, 37–38
    gain/system gain, 10, 17, 33, 34–35, 36,
        47, 49
    M2-F2 lifting body, 26, 27, 28–30, 86,
        91–92
    time lag between input and output, 34, 35,
        48
    31 X-15 research aircraft, 55, 56, 62–63,
        64
    X-31 and high AOA, 3
Cooley, David P. "Cools"
    A-LOC situation, 111
    age and G-LOC, 122
    anti-g straining maneuver, 111, 120–21
    education, training, and experience of,
        105–7, **106**, 121
    ejection from aircraft and windblast injuries,
        110, 111
    F-22A flight, 107–10, 120–23
Cords Road test area, 150–51
Corona satellite program, 61
Cotton, Joseph F. "Joe"
    B-52 pilot for M2-F2 launch, 85
    GE engine publicity photos, 132, 133, 134,
        135, 137, 140, 141
    XB-70A Valkyrie flight, 128–29

counter-pressure jerkin vest (COMBAT EDGE),
    119–20
coupling, multi-axis, 86
Crane, Carl, ix–x, 76
crew
    adaptability and flexibility, xiii
    atmospheric entry, g-forces, and positioning
        of, 23, 24
    crew-controller interactions, xiii
    qualifications of, xiv, xv, 190
crew resource management (CRM)
    B-1 accident, 155, **156**
    concept of, 20
    mishaps and, 189
    principles of, 20
    study and understanding of, xiii
    test flights, importance of during, 143
    X-31 accident and, 3, 16, 18–20, 21
Crippen, Robert, 50–51
Cross, Carl, 134, 135, 138
Crossfield, A. Scott, 56, 131
CSU-13B/P g-suit, 119
Cuddeback Dry Lake, 61–62
cultural issues, 159, 181–82
cupulogram (postrotatory nystagmus), 68,
    76–77

D

Dana, William, 57, 85
Defense Advanced Research Projects Agency
    (DARPA), 5, 6
Delamar Dry Lake, 63
delta-winged aircraft, 24, 128. *See also* XB-70A
    Valkyrie bomber prototype
Deutsche Aerospace/Messerschmitt-Bölkow-
    Blohm (MBB), 5
Dillon, Joseph F., 85
displays
    Attention-Based Principles, 74

design and function of, x–xi, 73–75, 78–79, 160, 189

Memory-Based Principles, 74–75

Mental Mode Principles, 74

multiple functions for, 63, 70–75, **70**, 103

Perceptual Principles, 74

distraction and attention, 94–98. *See also* attention

Doolittle, James H. "Jimmy," x, **x**

Dryden Flight Research Center/NASA Flight Research Center

depth-perception references on runways, 93–94

Haise and flight tests at, 41–42

lifting bodies program, 24–25. *See also* M2-F2 lifting body

Peterson's role at, 90

Scott as director of, 61

31 X-15 flight biomedical data collection, 59

X-31 flight test program, 5–7

See also Approach and Landing Test (ALT) program

Dutch roll motion, 87, 89, 93

Dyson, Norman K. "Ken," 5, 6

E

Edwards, George, 24

Edwards AFB, 133

Enterprise and ALT program, 36–37, 39, **40**

GE engine publicity photos, 132–41, **134, 136, 137, 138**

High-Speed Flight Station, 25

M2-F2 flight. *See* M2-F2 lifting body

Peterson's role at, 91

Reynolds's role at, 147

Ridley Mission Control Center, 149

X-15 program. *See* X-15 research aircraft

X-24B lifting body landing, 46

See also Dryden Flight Research Center/NASA Flight Research Center

Eggers, Alfred, 24

Eisenhower, Dwight, 127–28

ejection seats, 103, 110, 111, 145

elevon, 45–46, 48, 49

enabling factors, 180

engineering psychology. *See* human factors/human factors engineering

engines

GE engine publicity photos, **126**, 132–41, **133, 134, 136, 137, 138**

YF101-GE-100 turbofan engines, 144

YJ93 turbojet engines, 128, 139

Engle, Joe H., 38, 40

Enhanced Fighter Maneuverability (EFM) program, 2, 3, 5, 6. *See also* X-31 aircraft

Enterprise orbiter vehicle prototype

aerodynamic controls, 37–38

Approach and Landing Test Program, **32**, 37–38, **37**

autoland system, 38

captive flights, 38, 39

construction and materials, 37

crews and pilots for test flights, 38, **39**

flight controls, 44, 44n31

flight deck, 37

free flight and landing PIO incident, 43–46, **43**

free flights, 38, 39–41, **40**

landing speed, 38

PIO condition, operational fix for, 47–49

Shuttle Carrier Aircraft (SCA) launch, **37**, 38–40, 44

taxi tests, 38–39

visibility from cockpit, 46

environment and environmental constraints

errors and, xiii, **xiv**

Mir–Progress 234 collision, 184–85

study and understanding of, xiii

successful missions, systems and human factors for, 187, **187**

ergonomics. *See* human factors/human factors engineering

escape capsule, 145, **153**, 154–55, 157

Evenson, Mervin, 154

exercise protocols, 169–70, 174, 185

experimental aircraft, configuration changes to, 15–16

eyes and visual system

blackout, 101, 114–16, **115**

depth perception, 93–94

g-forces and, 114–16, **115**

grayout, 101, 112, 114–16, **115**

oculoagravic illusion, 55, 67–68, 69, 73, 77, 78

oculogravic illusion, 67

postrotatory nystagmus (cupulogram), 68, 76–77

visual-information processing, 95–97

F

F-4/F-4B Phantom, x, 132, 134, **134**

F-8 Digital Fly-By-Wire test bed, 47, 90

F-15 aircraft, 102, 104, 119

F-16 Fighting Falcon, 36, 107, 119

F-18 Hornet, **xii**, 8, 67

F-22 Combined Test Force (CTF), 105, 106, 107

F-22A Raptor

accident with, events of, 102, 107–10

accident with, factors in, 111, 120–23

Article 4002, **110**

Article 4008, **100**, 104–5, **108**

canopy, 103

chase planes, 104

cockpit, size of, 103

communications, 109, 111

construction and materials, 102, 104

delivery to Edwards AFB, 104

design and characteristics of, **100**, 102–3

ejection seat, 103, 110, 111

emergency systems, 103

flight controls, 103

flight tests, 104–5

g-suits/anti-g suits, 119, 120

glass cockpit, 103

pilot training, 107

refueling tanker, 104, 110

test profile and test-management plan, 122–23

weapons integration tests, 104–5, 107–10, **108**, **110**, 122

F-104 Starfighter

Adams and Scott landing accident, 60–61

design and characteristics of, 132

F-104D aircraft, 135

F-104G aircraft, 132

F-104N (NASA 813), 132

F-104N aircraft, cost of, 139

F-104N aircraft midair collision, 133–41, **134**, **136**, **137**, **138**

M2-F2 chase plane, **29**, 85, **87**

NASA 813 aircraft, 132

Walker flights, 130, 133–41

X-15 chase plane, **59**

XB-70 chase plane, 129, 131–32, 133, 139

F-111A jet aircraft, 84

F-117A aircraft, 107

F5D-1 aircraft, 25, 84, 85, **87**

fatigue

anti-g straining maneuver (AGSM), 118–19, 120–21, 122

anti-g suits and, 119, 120

asthenia and, 171

communications and, 18–19

g-tolerance and, 121–22

long-duration space travel and, 171

Mir–Progress 234 supply ship collision and, 159–60, 182–83

positive-pressure breathing (COMBAT EDGE), 119–20, 122

situational awareness and, 183

sleep deprivation and, 182–83
spatial disorientation (vertigo) and, 67–68
fighter aircraft
  g-forces and, 102, 121
  situational awareness and air combat
    maneuvering, 102
  See also specific aircraft
Fitts, Paul, x
flight controls and flight control systems (FCSes)
  acceleration and input, 34
  air-data failure, 11–12, 18, 19, 20
  air-data quality alerts, 9
  B-1 bomber, 146
  computerized FCSes and pilot workload, 99
  computerized FCSes and PIO, 36
  control gains adjustments, 10, 17, 33,
    34–35, 36, 47, 49
  control order, 34, 35
  design and location of controls, 16, 29–31,
    160, 189
  F-22A Raptor, 103
  flight performance and, ix
  fly-by-wire controls, 10, 36, 47
  g limit functions, 122
  gain/system gain, 10, 17, 33, 34–35, 36,
    47, 49
  input and loss of mode awareness, 55, 73,
    74–75, 77, 78
  input and PIO, 33–34, 35, 45–46
  MH-96 adaptive FCS, 55, 63, 64, **70**, 77–78
  pilot-induced oscillation, operational fix for,
    47–49
  pilot-induced oscillation and, 33–35
  reliability and principles of redundancy
    management, 10
  reversionary modes, 10–11, 12, 16, **17**
  simulator training and, 29–31
  Space Shuttle orbiter landing control, 49–51
  time lag between input and output, 34, 35,
    48

UAV controls, design and location of, 31
31 X-15 research aircraft, 55, 63, 64, **70**,
  77–78
X-31 aircraft, 5, 10–13, **10**, 20
flight instruments. *See* instruments and
  instrument panels
flight research and testing
  cognitive challenges during, ix
  flight conditions during, ix
  mission management, 143, 155–56
  people involved with, 143
  physiological challenges during, ix
  safety data and parameters, monitoring of,
    149–50
  test missions and "knock-it-off" option, 109
  test profile tasking (test card), 63, 146
Flight Research Center. *See* Dryden Flight
  Research Center/NASA Flight Research Center
Foale, Michael
  education, training, and experience of,
    165–66, **166**
  language skills of, 166
  maintenance and repairs, offer to perform,
    161–62, 166
  Mir, opinion of, **167**, 168
  Mir mission goals and experience, 170, 180
  Mir–Progress 234 supply ship collision, 178,
    179–80
  Tsibliyev's attitude toward, 161–62
Fraley, Stephen, 154
Fritz, John M., 132, 133, 134, 135, 137, 141
Fullerton, Charles Gordon
  Columbia mission, 51
  education, training, and experience of,
    42–43
  Enterprise and ALT program, 38, **39**, 40, 41,
    43, 44
Fulton, Fitzhugh "Fitz," 27, 38, 148

G

G Time-Tolerance curve (Stoll curve), 115–16, **115**, 118

g-forces

acceleration and, 101–2

age and G-LOC, 122

air combat maneuvering and, 102, 121

anti-g straining maneuver (AGSM), 111, 118–19, 120–21, 122

atmospheric entry, crew positioning, and, 23, 24

blunt-body aerodynamics, 24

cardiovascular system and, 112–13, **115**, 116, 121

centrifugal force, 101, 112, 113

centrifuge studies, 23, 115–16, 117, 122

cognitive function and, 102, 189

consciousness and, 101–2

FCSes and g limit functions, 122

g-induced loss of consciousness (G-LOC), 101–2, 112, 113, 114–18, **115**, 119, 121, 122, 189

g-warmup turns, 109

high-g test maneuvers with F-22A Raptor, 107–10, **110**

human tolerance for, 23

inertial force (+G$_z$), 101, 112, 121–22

lateral g-forces, 101, 112

physiological function and, 101–2, 111–23, 189

positive-pressure breathing (COMBAT EDGE), 119–20, 122

pulmonary system and, 112, 113–14

successive g exposures, 121–22

transverse g-forces, 101, 112

unit of acceleration forces (g), 112

visual system and, 114–16, **115**

g-suits/anti-g suits, 119, 120, 122

gain/system gain, 10, 17, 33, 34–35, 36, 47, 49

GBU-39 bombs, 105

Gemini program, 23, 61

General Electric (GE)

engine publicity photos, **126**, 132–41, **133**, **134**, **136**, **137**, **138**

YF101-GE-100 turbofan engines, 144

YJ93 turbojet engines, 128, 139

Gentry, Jerauld R., 82–83, 85

Gerathewohl, Siegfried, 67

Germany

EFC program, 5

X-31 test organization, 6

glass cockpits, xi, **xii**, 103

Glenn Research Center/Lewis Research Center, 41, 130

gravity and unit of acceleration forces (g), 112

Graybiel, Ashton, 67

grayout, 101, 112, 114–16, **115**

group dynamics, 159, 160, 161–62, 166, 170, 171–73, 175–76

Guidry, Thomas E., 38

H

H-21 helicopter, 85–86, 87–88, 94–95, 99

habit pattern transfer, 29–31, 91–92, 156–57

Haise, Fred W., Jr.

education, training, and experience of, 41–42

Enterprise and ALT program, 38, **39**, 40, 41, 42, 44–46

Hall, Michael, 88

Harmon International Trophy for Aviators, 131

Harper Dry Lake, 151

heat transfer, 58

helicopter, rescue, 85–86, 87–88, 94–95, 99

Henry, Stephen, 150

High-Speed Flight Station, 25

HL-10 lifting body, 80, 84–85

Hoag, Peter, 134

Honeywell

MH-96 adaptive FCS, 55, 63, 64, **70**, 77–78

X-31 flight control system, 5, 10–13, **10**, 20

Honts Trophy, 42, 60

Hook maneuver, 118. *See also* anti-g straining maneuver (AGSM)

Horton, Victor W. "Vic," 27, 38, 85

Human Factors Analysis and Classification System, **xiv**, 71, 182, 184, 189–90

human factors/human factors engineering

  birth of, x

  concept of, xii, xii n9

  configuration changes and, 15–16, 21

  controls, design and location of, 16, 29–31

  displays, design and function of, x–xi, 73–75, 78–79, 160, 189

  focus of and issues included in, xii

  pilot error, ix

  successful missions, systems and human factors for, 187, **187**

  "Swiss cheese" model of safety vulnerabilities, xiii, **xiv**, 127, 141, 159, 190

human-machine interaction

  accidents and mishaps and, xii

  computer technology and, xi–xii

  design of machine interface, 78, 189

  flight performance and, ix

  human performance and, x

  M2-F2 lifting body, 82–83

  Mir–Progress 234 collision, 185–86, **186**

  selection and training of human and, 78

  study and understanding of, xiii

  successful missions, systems and human factors for, 187, **187**

human performance

  confined environment and, 159, 171

  human-machine interaction and, x

  isolation and, 159, 171

Huxman, Joe, 88

I

inertial force, 101

input

  acceleration and, 34

  loss of mode awareness and, 55, 73, 74–75, 77, 78

  pilot-induced oscillation and, 33–34, 35, 45–46

  time lag between input and output, 34, 35

instability. *See* stability and instability

instrument flying, ix–x, 69, 76

instruments and instrument panels

  attitude indicator, 55, 63, 70–73, **70, 72**, 75

  confusing and difficult-to-read dials and switches, x–xi, **xi**

  controls, design and location of, 16, 29–31

  digital displays, 73–74

  displays, design and function of, x–xi, 73–75, 78–79, 160, 189

  displays, multiple functions for, 63, 70–75, **70**, 103

  glass cockpits, xi, **xii**

  safety and importance of, ix–x

  X-15 research aircraft, 62–63, 65, 69, 70–75, **70, 72**, 77–78

intercontinental ballistic missiles, 128

International Ergonomics Association, xii n9

International Space Station (ISS), 158, 160–61, 172–73

Isabella, Lake, 135, 141

K

Kaleri, Aleksandr, 175

KC-135R tanker, 104, 110

Kempel, Robert W., 89

Kennedy, John F., 128, 131

Kenyon, George, 24

Kerzon, Warren, 146

Kiel probe

  air data from, 7, 8, 9, 13, 16, 18, 19, 20, 21

configuration changes, awareness of, 14–16, 20–21

design and operation of, 7, **7**, 9, 13

icing conditions and, 7, 9, 13

no heat option, 8, 9, 13, 14–15, 16, 18, 20

Kim, Quirin, 7

Kincheloe Award, 131

King, Jay L., 88

Knight, William J. "Pete," 57, 62

knowledge-in-the-head, 74

knowledge-in-the-world, 74–75

Knox, Fred, 6

Korzun, Valeri, 170, 175

Kurs radar control system, 168, 177, 178–79, 182, 185

L

L-1 maneuver, 118. *See also* anti-g straining maneuver (AGSM)

Lacy, Clay, 133, 135, 136, 141

Lang, Karl-Heinz

character, experience, and professionalism of, 6–7, **6**

X-31 accident, 7–9, 14, 15, 16, 21

language barriers, 159

latent errors/failures, xiii, **xiv**, 190

Lazutkin, Aleksandr "Sasha"

education, training, and experience of, 163–64, **164**

group dynamics aboard Mir, 175–76

Mir–Progress 234 supply ship collision, 179–80

tasks aboard Mir, 163–64

Learjet, **133**, 134, 135

Lewis Research Center/Glenn Research Center, 41, 130

lift-to-drag ratio (L/D), 24, 47, 56, 58, **82**

lifting bodies, 23

blunt cone configuration, 24

blunt half-cone configuration, 24

design of and concepts behind, 23–25

development of, 24–25

landings and depth-perception references on runways, 93–94

lifting entry configuration, 24

spacecraft landing technique, development of, 23

X-15 flight data and design of, 58

Linenger, Jerry

communications, concerns about, 173

conditions aboard Mir, 174–76, **174**

education, training, and experience of, 164–65, **165**

group dynamics aboard Mir, 170, 175–76

language skills of, 173

Mir mission goals and experience, 164–65, 170, 180

Mir–Progress 233 docking problem, 162, 165

return from Mir, 165

LLRV (Lunar Landing Research Vehicle), 131

LN-3 navigational system, 132

Lockheed F-104 Starfighter. *See* F-104 Starfighter

Lockheed/Lockheed Martin F-22A Raptor. *See* F-22A Raptor

Lockheed YF-22 aircraft, 102

Loria, Gus, 7

Lousma, Jack, 51

Lovell, James, 42

Lucid, Shannon, 170

Lunar Landing Research Vehicle (LLRV), 131

M

M-1 maneuver, 118. *See also* anti-g straining maneuver (AGSM)

M2 lifting body, 23–25

M2-F1 lifting body, 26, 42, 80, 82, 84

M2-F2 lifting body

accident with, causes of, 91–99

accident with, events of, 83, 85–88, 90, **90**

accident with, inspection following, 88–89

aileron-rudder interconnection linkage, 26, 27, 28–30, 86, 91–92

B-52 mother ship launch, 26, 27–28, **27**, 84, 85–86

chase planes, **29**, 85, 86, **87**

communications, 87

construction and materials, 26

control surfaces adjustments, 26, 27, 28–30, 86

design and characteristics of, **22**, 26, 82–83, **82**

engines, 26

first flight, 23, **25**, 26–29

Gentry's flight, 82–83. *See also* Gentry, Jerauld R.

half-cone configuration, **22**

helicopter, rescue, 85–86, 87–88, 94–95, 99

human-system interface anthropometric standards, 82–83

landing gear, lowering of, 82–83, **82**, 87–88, 98–99

number of flights, 81

Peterson's flights, 80, 84. *See also* Peterson, Bruce A.

pilot-induced oscillation, 27, 28–29, 82, 89, 91–93, 94, 95, 98

pilot workload, 81

rockets, hydrogen-peroxide, 85

simulated landing flare, 26–27

simulator training, 29–31

stability of, 26, 86–87, 89

M2-F3 lifting body, 89

Man-in-Space Soonest project, 130–31

Manke, John A., 46, 85, 86, 87–88, 95

Manned Orbiting Laboratory (MOL), 42, 61, 77

March AFB, 88

Marine Corps, U.S., 117

Martin Company Moon-landing practice tests, 61

Martin Marietta, 102–3

Matejka, Robert, 68–69, 73

McCollom, John, 135, 140, 141

McLaughlin, Larry D., 85–86, 88

McMurtry, Thomas C., 38

medical screening tests, 68–69, 76–77

memory

asthenia and, 171

long-term, 97, 98

sensory, 97–98

short-term (working), 97, 98

Memory-Based Principles, 74–75

Mental Mode Principles, 74

Mercury program, 23, 131

Messerschmitt-Bölkow-Blohm (MBB)/Deutsche Aerospace, 5

MH-97 autopilot, 132

micromanagement, 170, 172–73

Mir space station

chain-of-command issues, 161, 181, 182

communications, 173, 175–76, 180–81, 184

conditions aboard, 166–68, **167**, 170, 174–76, **174**, 184

controls, design and location of, 160

crew arrivals and departures, 165, 166, 168

crew education, training, and experience, 162–66

cultural issues, 159, 181–82

design and characteristics of, **158**, 166–68, **167**

displays, design and function of, 160

docking devices and procedures, 160, 162, 163, 168, 177–79, **177**, **178**, 181–82, 184, 185–86, **186**

economic condition in Russia and, 160–61, 175

emergency systems, 184

environmental concerns, 184–85

exercise protocols, 169, 174, 185
extravehicular activities, 162, 163
fatigue and, 159–60
fire aboard, 164, 174–75, 176
funding for program, 160–61
group dynamics, 159, 160, 161–62, 166,
    170, 171–73, 175–76
human-machine interaction, 185–86, **186**
isolation feelings aboard, 170
language barriers, 159, 166, 170, 173
launch of, 160
maintenance and repairs, 161–62, 166, 175
micromanagement aboard, 170, 172–73
mission control, relationship with, 175–76,
    181
mission of, 160
NASA astronauts on, 158, 160, 161–62,
    164–66, **165**, **166**, **167**, 169–76, **174**,
    179–81
NPO Energia civilian cosmonaut on, 162,
    163–64, 182
NPO Energia ownership of, 161
oxygen-generating system, 174–75, 176,
    184–85
pilot workload and, 159–60
power supply and solar arrays, 179–80, **179**
Progress 233 near-accident, 160, 162, 163,
    165, 176, 177, 180, 181, 182
Progress 234 collision, causes of, 159,
    180–87, **188**
Progress 234 collision, damage from,
    179–80, **179**, 184
Progress 234 collision, events of, 159, 162,
    166, 177–80, **179**
situational awareness and docking attempt,
    180–81, 183, 184
sleep study and sleep deprivation, 176,
    182–83
Soyuz TM-17, strikes from, 163

system failures on, 159–60, 164–65, 166,
    174–76, 184–85
United States–Russian partnership, 160–61
workload and overscheduling, 170, 172–73
mishaps and accidents
causes of, xiv, **188**, 189–90
chain of events leading to, xiii–xv, 189–90
enabling factors, 180
human-machine interaction and, xii
iceberg analogy, **188**, 189
lessons learned from, xiv
organizational factors, xii, xiii, **xiv**, 180,
    189–90
pilot error as cause of, ix, xii
qualifications of people involved in, xiv, xv, 190
successful missions, systems and human
    factors for, 187, **187**
"Swiss cheese" model of safety
    vulnerabilities, xiii, **xiv**, 127, 141, 159,
    190
missile gap, 128
mission management, 143, 155–56
mission planning, xiii
Mk.82 bomb, 146, 150
mode awareness, 55, 73, 74–75, 77, 78
MOL (Manned Orbiting Laboratory), 42, 61, 77
Moon-landing practice tests, 61, 131
Munds, Frank, 135
muscle-mass losses in space, 169, 170–71
muscle memory, 29–31, 91–92
Myers, David, ix, 76

N

National Aeronautics and Space Administration
    (NASA)/National Advisory Committee for
    Aeronautics (NACA)
    blunt-body aerodynamics research by, 24
    creation of, 130–31
    GE engine publicity photos, 132–41, **134**,
        136, **137**, **138**

Lunar Landing Research Vehicle (LLRV), 131
missions, long-duration, 160–61
NASA 813 F-104 Starfighter, 132
NASA-Mir partnership, 160–61
Public Service Medal, 7
Russian Space Agency, communications
 with, 180–81
space exploration, responsibility for, 131
X-15 flight research program, 55, 131
X-31 program, 5
XB-70A Valkyrie program, 129
National Air and Space Museum, Smithsonian
 Institution, 58, 89
National Pilots Association, 131
Naval Safety Center, 189
Navy, U.S.
A-LOC, identification of, 117
A-LOC situations, 117
GE engine publicity photos, 132–33, 134
X-15 flight research program, 55
X-31 program, 5
neurovestibular function, ix, 66, 68–69, 76–77,
 93, 169, 170
North American Aviation
Apollo Command Service Module, 58
GE engine publicity photos, 132–41, **133**,
 134, 136, 137, 138
Rockwell International merger, 58
X-15 research aircraft, 55, 56, 58
XB-70A Valkyrie bomber prototype, 128
Northrop Aircraft Corporation, 83, 88, 89, 90–91
Northrup Strip (White Sands Space Harbor), 43
NPO Energia, 182
civilian cosmonaut aboard Mir, 162, 163–64
ownership of Mir, 161

O

Ocker, Bill, ix–x, 76
Octave Chanute Award, 131, 148
oculoagravic illusion, 55, 67–68, 69, 73, 77, 78

oculogravic illusion, 67
organizational factors
climate, organizational, 159, 186, 190
communications philosophy, 18, 190
mishaps and, xii, xiii, **xiv**, 180, 189–90
mission outcome, people involved in, 127
policy compliance and operational
 procedures, xii, xiii, **xiv**, 132–33, 140–41,
 159, 189
resource management, 159
small groups and organizational behavior, 20
successful missions, systems and human
 factors for, 187, **187**
oscillation
pitch oscillation, 64
See also pilot-induced oscillation (PIO)

P

parabolic space flight profile, 55, 67, 79
Paraglider Research Vehicle (Paresev) program,
 26, 83–84
Perceptual Principles, 74
Peterson, Bruce A.
education, training, and experience of, **80**,
 82, 83–85, 89–91
injuries to, 88, 89
M2-F2 accident, 83, 85–88, 90, **90**, 91–95
M2-F2 flights, 80, 84
spatial disorientation (vertigo), 92–94
physiological function
almost-loss of consciousness (A-LOC), 102,
 111, 117–18, 121, 189
anti-g straining maneuver (AGSM), 111,
 118–19, 120–21, 122
asthenia, 171
cardiovascular system and g-forces,
 112–13, **115**, 116, 121
flight tests and challenges to, ix
g-forces and, 101–2, 111–23, 189

g-induced loss of consciousness (G-LOC), 101, 112, 113, 114–18, **115**, 122

incapacitation due to G-LOC, 116, 121, 189

long-duration space travel, 169, 170–71

medical screening tests, 68–69, 76–77

muscle-mass and bone-density losses, 169, 170–71

neurovestibular function, ix, 66, 68–69, 76–77, 93, 169, 170

oculoagravic illusion, 55, 67–68, 69, 73, 77, 78

oculogravic illusion, 67

pulmonary system and g-forces, 112, 113–14

spatial orientation in flight, research on, ix–x

study and understanding of, xiii, 189

successive g exposures, 121–22

vertigo (spatial disorientation), 55, 62, 65–70, 76–77, 92–94

visual system and g-forces, 114–16, **115**

X-15 flight biomedical data collection, 59

pilot-augmented oscillations, 35. *See also* pilot-induced oscillation (PIO)

pilot error

Cambridge Cockpit research on, x

hardware interface and, x

mishaps from, ix, xii

pilot-in-the-loop oscillations, 35. *See also* pilot-induced oscillation (PIO)

pilot-induced oscillation (PIO)

additional names for, 35

categories of, 36

definition of, 33

elements required to produce, 33–35

Enterprise free flight and landing PIO incident, 43–46, **43**

Enterprise PIO condition, operational fix for, 47–49

first incident of, 33

fly-by-wire controls and, 36

input and, 33–34, 35, 45–46

lateral oscillation, 89, 92–93, 94, 95, 98

M2-F2 lifting body, 27, 28–29, 82, 89, 91–93, 94, 95, 98

as pejorative term, 35

pilot role in, 35

reduction or elimination solutions, 35, 47–51

severity of, 35–36

simulated landing flare to evaluate, 27

time lag between input and output, 34, 35

training to reduce or eliminate, 35, 48–49, 51

pilot workload

attention and, 98–99

computerized FCSes and, 99

control order and cognitive workload, 34

dials and switches, confusing and difficult-to-read, **xi**, 81

formation flying and, 139–40

M2-F2 accident, 81

Mir–Progress 234 supply ship collision and, 159–60

Mir workload and overscheduling, 170, 172–73

performance and task saturation, 81

situational awareness and, 93

spare capacity, 81

task saturation, x–xi, 81, 99

X-15 accident and, 65, 71

X-15 electrical malfunction and, 63, 66

X-15 FCS and, 63

X-31 accident and, 21

pilots

astronaut wings earned on X-15 flights, 56–57, 62, 65

bored with flying, 81

characteristics of, ix

cowboy image of, **viii**

education and training of, ix

hardware interface and flight performance, x

medical screening tests, 68–69, 76–77

PIO, training to reduce or eliminate, 35, 48–49, 51

qualifications of, xiv, xv, 190

skill of and flight performance, ix

pitch-up illusion, acceleration and, 66–67

pitot tube

configuration changes, awareness of, 14–16

heating of and heat switch, 8, 9, 12, 13, 14, 15, 16, 18, 20

icing conditions and, 8, 9, 12, 13, 14

replacement of with Kiel probe, 7, 9

Rosemount probe, 9, 12, 13, 15

signal from and air-data failure, 11–12

Polyakov, Valery, 176

positive-pressure breathing (COMBAT EDGE), 119–20, 122

postrotatory nystagmus (cupulogram), 68, 76–77

preconditions for unsafe acts, 182, 184

pressure suits

altitude requirements for, 25

M2-F2 flight, no requirement for, 25, 27

Progress-M spacecraft

design and characteristics of, 168, **168**

Mir, docking with, 168

Progress 233 near-accident, 160, 162, 163, 165, 176, 177, 180, 181, 182

Progress 234 collision with Mir, 159, 162, 166, 177–80, **179**

Progress 234 collision with Mir, causes of, 159, 180–87, **188**

psychological factors, xiii, 160, 171, 172, 176

pulmonary system and g-forces, 112, 113–14

Purifoy, Dana, 8

R

reaction control system and hydrogen-peroxide-fueled thrusters, 55, 56, 63

Reagan, Ronald, 145

Reason, James, xiii–xiv, 127, 141, 159, 190

record flights

altitude, 56, 57, 129, 131

speed, 56, 57, 131, 145

Reed, Robert Dale, 24

Reynolds, Richard V.

B-1 bomber Flight 2-127 test flight, 146–47, 150–57

education, training, and experience of, **146**, 147

Richter, Helmut, 6

Ridley Mission Control Center, 149

Rockwell International

B-1 bomber program, 143. *See also* B-1 bomber

B-1 Combined Test Force, 145

Enterprise, 37. *See also* Enterprise orbiter vehicle prototype

North American Aviation merger, 58

Temporary Operating Procedure (TOP) system, 14

X-31 program, 5, 6

Rogallo Paraglider Research Vehicle (Paresev) program, 26, 83–84

Rogers Dry Lake

characteristics of, 94

Columbia landing, 50–51

depth-perception references, 93–94

M2-F2 program, 26, 27, **29**, 87, **87**, 88, **90**, 94

Space Shuttle orbital flight-test mission, 43

X-15 landings, 56, **59**

rolling motions

Dutch roll motion, 86, 89, 93

M2-F2 lifting body, 86–87, 89, 91–94

Roman, James, 92

Royal Australian Air Force, 118

rudder-aileron interconnection linkage, 26, 27, 28–30, 86, 91–92

runways

concrete runway landings, 46
Enterprise landing and PIO incident, 43–46, **43**
precise landings, 46
Russia
economic condition in, and Mir, 160–61, 175
International Space Station agreement, 160–61
Ukraine, relationship with, 168
Russian Space Agency (RSA)
chain-of-command issues, 182
military control and oversight, 181
Mir and mission control, relationship between, 175–76, 181
Mir fire and public relations, 175
NASA, communications with, 180–81
organizational culture, 186
personnel pay, 161, 164, 181

S

safety
preconditions for unsafe acts, 182, 184
"Swiss cheese" model of safety vulnerabilities, xiii, **xiv**, 127, 141, 159, 190
systems and human factors for successful missions, 187, **187**
unsafe acts, 182, 184
Saturn V rocket, 58
SCAS (stability control augmentation system), 153
Scott, Dave, 60–61
Seeck, Dietrich, 5, 6
Shackelford, Linda, 169
Shepard, Van, 129
Short Range Attack Missile, 146
silent incapacitation, 155
situational awareness
air combat maneuvering and, 102

almost-loss of consciousness (A-LOC) and, 102, 111, 117–18, 121, 123, 189
attention and, 98
F-22A Raptor accident and, 102, 111, 117–18, 123
fatigue and, 183
Mir–Progress 234 collision, 180–81, 183
mishaps and, xiii, **xiv**
mode awareness, 55, 73, 74–75, 77, 78
pilot workload and, 93
spatial disorientation (loss of situational awareness), 101–2, 117, 123
spatial orientation and, 102
successful missions, systems and human factors for, 187, **187**
task saturation and, 93
X-31 accident and, 3, 19–20
Six Million Dollar Man, The, 90
Skylab, 160, 172
Skyrud, Jerome P., 134, 135
sleep study and sleep deprivation, 176, 182–83
Smith, James G., 133, 140
Smithsonian Institution, National Air and Space Museum, 58, 89
Society of Experimental Test Pilots, 131, 148
somatogravic illusion, 66–67, 69
Sorlie, Donald M., 82, 85
Soviet Union
dissolution of, 160
missile gap, 128
Soyuz spaceships
Mir, docking with, 168
Mir crew arrival and departure by, 168
Soyuz TM-17 spacecraft, 162–63
space flight, long-duration
activities and duties, 172–73
asthenia, 171
bone-density losses, 169
exercise protocols during, 169–70, 174, 185
isolation feelings, 159, 170, 171

muscle-mass losses, 169, 170–71
NASA corporate knowledge about, 160–61
systems and human factors for successful
   missions, 187, **187**
Space Shuttle missions
   activities and duties on, 172–73
   Fullerton's missions, 43
   length of, 172, 173
   Mir crew arrivals and departures, 165, 168
   Space Transportation System (STS)-51F
     mission, 43
   Space Transportation System (STS)-63
     mission, 166
Space Shuttle orbiter
   autoland system, 38, 51
   cockpit location, 38, 48, 49, 51
   design and characteristics of, 36, 38, 56, 58
   handling difficulties, 38
   landing of, 49–51
   lifting entry configuration, 24
   low-speed handling of, 36
   PIO, modifications to reduce or eliminate,
     47–51
   visibility from cockpit, 46, 48
   windshield design, 58
   X-15 flight data and design of, 56, 58
   See also Approach and Landing Test (ALT)
     Program
spacecraft
   ballistic entry and crew positioning, 23–24
   entry into the atmosphere and crew
     positioning, 23–24
   landing technique, lifting body program to
     develop, 23
   lifting entry configurations, 24
   X-15 flight data and design of, 58
Spartan satellite, 166
spatial disorientation
   loss of situational awareness, 101–2, 117,
     123

   vertigo, 55, 62, 65–70, 76–77, 92–94
spatial orientation, ix–x, 55, 102
speed
   low-speed flight and maneuvers, ix, 36, 143,
     144, 146, 151–52
   record flights, 56, 57, 131, 145
   X-15 high-speed flights, 56, 57
spin and spin recovery
   flat spin, 64
   X-15 research aircraft, 64, 78
SST (supersonic transport) program, 128, 129,
   134, 148
stability and instability
   B-1 and wing-sweep maneuvers, 144, 149,
     151–53, **151**, 155–56, 157
   closed-loop instability, 33, 35, 89
   gain and, 34–35
   of M2-F2 lifting body, 26, 86–87, 89
stability control augmentation system (SCAS),
   153
Stallings, Herbert, 67
Stoll, Alice, 115
Stoll curve (G Time-Tolerance curve), 115–16,
   115, 118
Strategic Air Command, 145
Sturmthal, Emil "Ted," 145
supersonic transport (SST) program, 128, 129,
   134, 148
Swigert, John, 42
"Swiss cheese" model of safety vulnerabilities,
   xiii, **xiv**, 127, 141, 159, 190
Syvertson, Clarence, 24

T

T-38 aircraft, 62, 85, 132, 134, **134**
task saturation
   attention and distraction, 94–98
   pilot workload, x–xi, 81, 99
   situational awareness and, 93
   spatial disorientation (vertigo) and, 67–68

Telerobotically Operated Rendezvous Unit
(TORU), 163, 168, 177–79, **177**, 182, 184

Tenerife, Spain, 19

test missions and "knock-it-off" option, 109

Test Pilot School, Air Force, 84, 107

test profile tasking (test card), 63, 146

Thagard, Norman, 169, 170

Thompson, Milton O. "Milt"

    bored with flying, 81

    experience and training of, 25–26, **25**

    M2-F2 lifting body flights, **25**, 27–29, **27**,
    81–82

    simulator training, 29–31

    Walker, opinion about, 130

Three Sisters Dry Lake, 135, 141

thrust vectoring, 3–5, **4**, 10–11

thrusters, hydrogen-peroxide, 55, 56

TIFS (Total In-Flight Simulator), 47–49

time lag, 34, 35, 48

TORU (Telerobotically Operated Rendezvous
Unit), 163, 168, 177–79, **177**, 182, 184

Total In-Flight Simulator (TIFS), 47–49

Trippensee, Gary, 6–7

Truly, Richard H., 38, 40

Tsibliyev, Vasily

    education, training, and experience of,
    162–63, **162**, 184

    Foale, attitude toward, 161–62

    group dynamics aboard Mir, 175–76

    health of, 186

    Mir docking device and procedures, 163,
    181–82, 184, 185–86

    Mir–Progress 233 docking problem, 176,
    177, 182

    Mir–Progress 234 collision, 177–80, **178**,
    184

    situational awareness, 180, 184

    sleep study and sleep deprivation, 176,
    182–83

    Soyuz TM-17 flight, 162–63

temper of, 175

U

Ukraine

    docking devices from, 160, 168

    economic condition in and Mir, 160

    Russia, relationship with, 168

United States (U.S.)

    economic condition in, 160–61

    International Space Station agreement, 158,
    160–61, 172–73

    missile gap, 128

unmanned aerial vehicles (UAVs), 31

unsafe acts, 182, 184

V

velocity

    control order of systems, 34

    X-15 flight and sense of, 77

vertigo (spatial disorientation), 55, 62, 65–70,
76–77, 92–94

virtual reality systems, 169–70

visual system. *See* eyes and visual system

W

Walker, Joseph A. "Joe"

    astronaut wings earned on X-15 flights, 57,
    131

    death of, 136

    education, training, and experience of, **viii**,
    129–32, **129**, 139

    F-104 chase plane flights, 129, 131–32,
    133, 139

    GE engine publicity photos and midair
    collision, 133–41, **134**, **136**, **137**, **138**

    Thompson's opinion of, 130

    X-1A test flight, **viii**

    X-15 flights, 56, 131

    XB-70A Valkyrie flight, 129

Waniczek, Otto J.

B-1 bomber Flight 2-127 test flight, 146–47, 150–57

education, training, and experience of, 149

weapons integration tests

F-22A Raptor, 104–5, 107–10, **108**, **110**, 122

F-117A aircraft, 107

White, Al, 128–29, 134, 135, 136, 137–38

White, Robert A., 56

White Sands Missile Range, 43, 51

White Sands Space Harbor (Northrup Strip), 43

windshield design and bimetallic "floating retainer" concept, 58

Wolfe, Tom, ix

workload. *See* pilot workload

Wright, Orville, 33

Wright, Wilbur, 33

Wright Flyer, 33

Wright-Patterson AFB, 42, 57

X

X-1/X-1A aircraft, **viii**, ix, 130

X-15 research aircraft

acceleration and pitch-up illusion, 66–67

accident with X-15-3, causes of, 55, 65–75, 77–79

accident with X-15-3, events of, 55, 62–65, **65**, 68–69, 70–73, **72**

accident with X-15-3, re-creation of, **78**

accidents with, 57

astronaut wings earned on, 56–57, 62, 65, 131

attitude indicator, 55, 63, 70–73, **70**, **72**, 75

B-52 mother ship launch, 27, 56, 63

biomedical data collection, 59

contractor, 55

control surfaces and aerodynamic control, 55, 56, 62–63, 64

data from flights, aircraft and spacecraft development from, 56, 58

design and characteristics of, **54**, 55, 131

difficulties and glitches during flights, 62

electrical malfunction, 63, 64, 66, 77–78

flight control system, 55, 63, 64, **70**, 77–78

flights with, 56–58, 131

high-altitude flights, 56, 57

high-speed flights, 56, 57

instruments and instrument panels, 62–63, 65, 69, 70–75, **70**, **72**, 77–78

landings, 56, 58, **59**

landings and depth-perception references on runways, 93–94

number of flights, 55–56

purpose of, 54, 55

reaction control system and hydrogen-peroxide-fueled thrusters, 55, 56, 63

record flights, 56, 57, 131

reentry, aerodynamic heating, and thermal protection, 58

research missions, 56

roll indication, 55, 63, 70–71, 73

speed brakes, insulation on, 58

spin and spin recovery, 64, 78

test profile tasking (test card), 63

Thompson as pilot, 26

thrusters, hydrogen-peroxide, 55, 56

vertigo (spatial disorientation), 55, 62, 65–70

weight of, 66

windshield design and bimetallic "floating retainer" concept, 58

X-15-1 aircraft, 63

X-15-2 aircraft, 63

X-15-3, 62–65, **65**

X-15A-2 aircraft, 57

XLR99 rocket engines, 66

yaw indication, 55, 63, 70–71, 73

X-20 Dyna-Soar spaceplane, 24, 26, 148

X-24B lifting body, 46

X-31 aircraft

accident with, causes of, 3, 7–21

accident with, events of, 8–9, 16, **19**

automation bias and, 3, 10–13

communications and, 3, 7–8, 9, 14–15, 16, 18–21

configuration changes, awareness of, 13–16, 20–21

construction and materials, 5

CRM (crew resource management) and, 3, 16, 18–20, 21

design of, 5, 21

flight control system, 5, 10–13, **10**, 16, **17**, 20

flight test program, 5–7

goal of research program, 3, 5

high-AOA (angle of attack) capabilities, 3–6, **4**

icing conditions and, 8, 9, 12, 13–15, 16, 18

Kiel probe and air data. *See* Kiel probe

photos of, **2**, **4**

pilots for, 5, 6, 7–8

pitot tube. *See* pitot tube

situational awareness and, 3, 19–20

system-safety analysis, 11–12, 13

test organization, 6

X-planes, ix

XB-70A Valkyrie bomber prototype

Air Vehicle Two (AV-2), 128–29

chase planes, 129, 131–32, 133, 139

cost of, 139

design and characteristics of, **126**, 128

ejection from, 137–38

engines, 128, 139

first flights, 128–29

flight tests, 128–29, 131–32, 133–34, 135

focus of program, 128

funding for program, 127–28

GE engine publicity photos and midair collision, 132–41, **134**, **136**, **137**, **138**

number built, 127, 128

record flights, 129

schedule for development, 127

SST program and, 128, 129, 134

wave-rider concept, 128

XLR11 rocket engines, 26

XLR99 rocket engines, 66

Y

Yeager, Charles E. "Chuck," ix, xii, 60

YF-5A aircraft, 133, 134, **134**

YF-16A aircraft, 36

YF101-GE-100 turbofan engines, 144

YJ93 turbojet engines, 128, 139

Young, John, 50

Young, William R., 38

Z

Zimmerman, Keith, 162

CPSIA information can be obtained
at www.ICGtesting.com
Printed in the USA
LVHW080124241122
733952LV00005B/57

9 781782 662464